SpringerBriefs in Electrical and Computer Engineering

W0017246

For further volumes:
http://www.springer.com/series/10059

Ilya Gertsbakh · Yoseph Shpungin
Radislav Vaisman

Ternary Networks

Reliability and Monte Carlo

 Springer

Ilya Gertsbakh
Department of Mathematics
Ben Gurion University
Beer-Sheva
Israel

Radislav Vaisman
School of Mathematics and Physics
University of Queensland
Brisbane, QLD
Australia

Yoseph Shpungin
Department of Software Engineering
Shamoon College of Engineering
Beer-Sheva
Israel

ISSN 2191-8112   ISSN 2191-8120   (electronic)
ISBN 978-3-319-06439-0   ISBN 978-3-319-06440-6   (eBook)
DOI 10.1007/978-3-319-06440-6
Springer Cham Heidelberg New York Dordrecht London

Library of Congress Control Number: 2014937273

Printed on acid-free paper

Springer is part of Springer Science+Business Media (www.springer.com)

Preface

Traditionally, Reliability theory studies networks whose components subject to failure (edges and/or nodes) are binary, i.e., have two states—*up and down*. Typically, network *DOWN* state is defined as loss of terminal connectivity or a break-up into a critical number of components. There are two most often used tools for investigating network reliability—Monte Carlo simulation and the so-called D-spectra technique.

The "standard" model in network Monte Carlo [1] assumes that components fail independently, and node (or edge) i fails with probability q_i. The goal of Monte Carlo simulation is estimation of network static *DOWN* probability. In our opinion, the most efficient and accurate Monte Carlo method is the so-called evolution and merging algorithm originally suggested by M. V. Lomonosov [2, 3]. It has been shown [4] that this algorithm can be easily adapted to the case of non-reliable nodes. When all components are statistically independent and identical, i.e., have the same *down* probability q, reliability analysis can be considerably simplified by using the so-called D-spectra or signature technique. D-spectrum is a discrete distribution $\mathbf{f} = (f_1, f_2, \ldots, f_n)$, where f_i is the probability that system failure takes places at the instant of the i-th component failure. D-spectrum is system combinatorial invariant. It depends only on system structure function and has the following surprising property: the number $C(x)$ of system failure sets with exactly x components *down*, can be expressed via the D-spectrum by the following simple formula:

$$C(x) = (f_1 + \cdots + f_x) \cdot n!/(x!(n-x)!)$$

The cumulative D-spectrum $F(x) = f_1 + \cdots + f_x$ can be easily estimated by means of Monte Carlo simulation. To calculate $P(DOWN)$ we use the formula:

$$P(DOWN) = \sum_{j=1}^{n} C(j)q^j p^{n-j} . (*)$$

D-spectra technique works quite well for networks of small to medium size with 30–100 components. All D-spectra-based techniques can be easily extended to the binary networks having more than two states [5]. Moreover, a modification of D-spectra allows obtaining another system invariant, so-called Importance Spectrum, by means of which it becomes possible to calculate component Birnbaum Importance Measure (BIM), widely used in network design [2, 3, 6–8].

Summing up, network reliability theory successfully handles systems with two principal limitations:

(i) components fail independently;
(ii) components are binary.

We do not see an easy way to relax assumption (i), except for some *ad hoc* situations. Since violation of independence cannot be formalized into a simple "dependence" model, we cannot expect, in our opinion, rapid and decisive progress in this direction.

What about the second assumption? Almost every practical application of networks to real-life situations puts a question mark to the binary assumption regarding the component state. Here are some examples.

If the network describes the road system, link $e = (a, b)$ failure means traffic violation between nodes a and b. This violation almost never means complete disruption of transport flow. Natural disaster like flood or earthquake, may lead to only partial damage of the road segment. Therefore, between the perfect state of a link and its complete failure there should be at least one intermediate state.

Similar is the situation in flow networks. If an edge $e = (a, b)$ representing water supply pipe is in perfect state (*up*), it can deliver maximal amount of water, say of 1,000 cube/h. If it is broken (*down*) no water is delivered from a to b. But there are also situations when due to some technical reasons (e.g., leaks, maintenance works, partial damage) the water flow is reduced by 50 or 30 %. This means that an adequate description of the link asks for introducing one or several intermediate states between *up* and *down*. Communication networks consist of communication lines (channels) allowing different rates of information transmission depending on the technical state of the channel. In the absence of interferences (*up* state) the transmission speed is maximal. In case of broken channel (*down*) the speed is reduced to zero. It may happen that during peak hours, the transmission speed gets reduced by 30 %, which represents an intermediate state "between" *up* and *down*.

Social networks representing connections between individuals have also several degrees of "closeness" between two individuals who either maintain an extensive information exchange, or have no exchange at all, or have a reduced level of communication.

This book is devoted to the reliability analysis of ternary (or trinary) networks, i.e., to networks whose components subject to failure have *three* states: *up, down* and an intermediate state which we call *mid*. It turns out that the D-spectra technique can be extended to the case of components with three states. The price for this extension is introducing a more complicated version of D-spectrum, the *ternary D-spectrum*.

Ternary D-spectrum is a collection of so-called cumulative r-spectra $F_r(x)$, $r = 0, 1, \ldots, n - 1; x = 1, 2, \ldots, n - r$. Here $F_r(x)$ is the probability that the network is *DOWN* if r of its components are *up*, x components are *down* and the remaining $(n - r - x)$ are in state *mid*. Formally, instead of a vector **f** in binary case, we have a set of vectors for ternary case. The approximate computation of the ternary spectrum is carried out by a quite straightforward Monte Carlo procedure.

Ternary spectrum is a network combinatorial invariant, and if it is known in addition to the fact that all components are statistically independent and identical, system *DOWN* probability can be computed by means of a simple formula similar to (*). The material of this book is organized as follows.

Section 1.1–1.4 of Chap. 1 are devoted to the definition and properties of the ternary D-spectrum. The knowledge of the ternary D-spectrum $\{F_r(x)\}$ allows counting the network failure sets with given structure. Let $C(r; x)$ be the number of failure sets having r components *up*, x components *down* and the remaining $n - r - x$ components in *mid* state. Then we prove that

$$C(r; x) = F_r(x) \cdot \frac{n!}{r! x! (n - r - x)!} \cdot (**)$$

If all network components are independent and identical, (**) allows to find network *DOWN* probability by means of the following formula:

$$P(DOWN) = \sum_{\{r, x : 0 \le r + x \le n\}} C(r; x) p_2^r p_1^{(n-r-x)} p_0^x,$$

where p_2, p_1, p_0 are components *up*, *mid*, and *down* probability, respectively.

Section 1.5 of Chap. 1 is devoted to a modification of ternary D-spectrum called ternary importance spectrum which allows to evaluate network component importance measures. These measures are a modification of Birnbaum Importance Measures [3, 5–8] adjusted to ternary components.

Section 1.6 describes how to obtain an approximation to the ternary D-spectrum and to the component importance measures using Monte Carlo simulation techniques.

Chapter 2 consists of two parts. The first part (Sects. 2.1–2.4) is a numerical illustration of the theory developed in Chap. 1. In Sect. 2.1 we consider reliability calculations for an H_4 network. The network is a hypercube of order four, it has 16 nodes and 32 edges. Next, we define in it two sets of terminals, T_1 and T_2. An edge $e = (a, b)$ in state *up* provides high communication speed between a and b. If this edge is in state *mid*, the $a \leftrightarrow b$ communication goes with reduced speed; *down* state for an edge means that this edge does not exist. Edge state is chosen randomly and independently, according to probabilities p_2, p_1, p_0 for *up*, *mid*, and *down* state, respectively. System *UP* state is defined as the existence of high-speed communication between nodes of T_1 *and* the existence of a path of operational edges between any pair of nodes of T_2. We present data on network reliability and on the ternary D-spectrum.

Section 2.2 considers a stochastic source—terminal problem for a dodecahedron network. In this network, an edge $e = (a, b)$, except for edges going out of s and into t, is in fact a *pair* of directed links for $a \to b$ and $b \to a$ directed flows. Each link has capacity 6, 3, or 0 for *up*, *mid* and *down* state, respectively. The network has two *DOWN* states, *DOWN2* and *DOWN1*, for the flow less than

L_1 or L_2, respectively ($L_2 < L_1$). We present data on network reliability for various values of edge probability vectors $p = (p_2, p_1, p_0)$.

Section 2.3 is an example of a rectangular grid network with 100 nodes and 180 edges. Components subject to failure are the nodes. If a node is *down* all edges adjacent to it are erased and the node gets isolated. If a node is in *mid* state, it has only horizontal or vertical edges, depending on the position of the node. For this network we calculate the probability that the largest connected node set (an analogue to a "giant" component) has less than L nodes.

The first part of Chap. 2 is concluded by Sect. 2.4 which presents edge importance data and their analysis for H_4 network. The second part (Sect. 2.5) deals with networks which have statistically independent and *nonidentical* components. Component i has state distribution $p^{(i)} = (p_2^{(i)}, p_1^{(i)}, p_0^{(i)})$ meaning that the component is in state *up*, *mid*, and *down* with probability $p_2^{(i)}, p_1^{(i)}$ and $p_0^{(i)}$, respectively. In this situation, different failure sets with the same number of components in *up, mid, down* have different probabilistic weights, and this makes it not possible to use the ternary spectrum technique for finding system *DOWN* probability. What remains in this more complex case is to resort to a fast and accurate Monte Carlo method. Such a method is based on a modification of M. V. Lomonosov's evolution algorithm [2, 3]. The algorithm is described in Sects. 2.5.1 and 2.5.2. Its action is illustrated by numerical examples of flow and grid networks.

In reality, networks usually interact with each other and failure in one network causes failure in another one. For example, functioning of a city road network strongly depends on the traffic light power supply system: several non-functioning traffic lights ("nodes") may cause traffic jams in large areas. Another example is power supply network and communication network which strongly interact with each other.

The simplest form of two interacting networks is sharing the same set of nodes by two independent networks. For example, the power supply and water supply networks in the same geographic area share the same set of nodes (houses or residencies). Section 3.1 presents several simple results concerning the size of the set of nodes which receive "full" supply, i.e., are adjacent to edges of both types. Here we use some basic facts from the theory of large random Poisson networks [9].

Section 3.2 considers a system of two or more finite interacting networks. Here the interaction means that a node v_a of network A delivers "infection" to a randomly chosen node v_b in B which in turn, bounces back and infects another randomly chosen node w_a in network A, and so on. As a result, a random number Y of nodes in B gets "infected" and fails. We compute, using D-spectra technique, the *DOWN* probability for network B. This model is generalized to the case of several peripheral networks attacking one "central" binary network B. In this "attack," some nodes in B will receive more than one hit. The use of DeMoivre combinatorial formula combined with the D-spectra technique allows us to obtain in a closed form an expression for network B *DOWN* probability.

Finally, Sect. 3.3 extends the results of Sect. 3.2 to the case when the "central" network is ternary. In that case, we must take into account that nodes which were hit once or more will be in different states. It is assumed that a node hit only once changes its state from *up* to *mid*. When this node receives another hit, it turns into *down* and remains in it forever. Network *DOWN* probability for this case can be estimated by a Monte Carlo algorithm.

George Box used to cite the aphorism: "all models are wrong; some models are useful." We hope that some models presented in this book might be useful to reliability researchers involved in network study and design, and to reliability engineers interested in applications of the theory to practical calculations of network reliability parameters.

Acknowledgments

My work was partially supported by the Australian Research Council under grant number CE140100049.

March 2014

Ilya Gertsbakh
Yoseph Shpungin
Radislav Vaisman

References

1. Kroese D, Taimre T, Botev ZI (2011) Handbook of Monte Carlo methods, Chap. 16. Wiley, New York, pp. 549–576
2. Elperin T, Gertsbakh IB, Lomonosov M (1991) Estimation of network reliability using graph evolution models. IEEE Trans Reliab 40(5):572–581
3. Gertsbakh I, Shpungin Y (2009) Models of network reliability: analysis, combinatorics and Monte Carlo. CRC Press, Boca Raton
4. Gertsbakh I, Shpungin Y, Vaisman R (2014) Network reliability Monte Carlo with nodes subject to failure. Int J Perform Eng 10(2) 161–170
5. Gertsbakh I, Shpungin Y (2011) Network reliability and resilience, Springer Briefs in Electrical and Computer Engineering, Springer. Berlin, Heidelberg
6. Barlow RE, Frank Proschan (1975) Statistical theory of reliability and life testing. Holt, Rinehart and Winston, Inc., New York
7. Birnbaum ZW (1969) On the importance of different components in multicomponent system. In: Krishnaiah PR (ed) Multivarite analysis-II. Academic Press, New York, pp 581–592
8. Gertsbakh I, Shpungin Y (2012) Combinatorial approach to computing importance indices of coherent systems. Probab Eng Inf Sci 26:117–128
9. Newman MEJ (2010) Networks. An introduction. Oxford University Press, Oxford

Contents

Notations

$up, down, mid$	Component states in ternary system
c.d.f.	CDF—cumulative distribution function
τ, X, Y, Z	Random variables
$\chi(t)$	Time-continuous random process
$X \sim Exp(\lambda)$	r.v. X is exponentially distributed with parameter λ
CPU time	Central process unit computation time
RE	Relative error
$UP, DOWN$	System operational and failure state, respectively
$\mathcal{N} = (V, E, T)$	Network with node (vertex) set V, edge (link) set E and terminal set T
p_2, p_1, p_0	Probability that a component is in state up, mid ($middle$) and $down$, respectively
$\mathbf{y} = (y_1, y_2, \ldots, y_n)$	System (network) state vector; $y_i = 2$, $y_i = 1$, or $y_i = 0$ if component i is up, mid, or $down$, respectively
$\pi = \left(i_1^{(2)}, \ldots i_r^{(2)}; i_{r+1}^{(1)}, \ldots, i_n^{(1)} \right)$	r-type random permutation of component numbers, in which on the first r positions the components are in up, and on the remaining—in mid
$\{F_r(x), r = 0, \ldots, n-1; x = 0, \ldots, n-r\}$	Ternary D-spectrum. r is the number of components in up, x—the number of components in $down$

$C(r; x)$ Number of system failure sets having r components *up* and x components *down*

$\tau \sim Exp(\Lambda)$ Random variable τ is exponentially distributed with parameter Λ

Chapter 1
Networks with Ternary Components: Ternary Spectrum

Abstract In this chapter we consider a monotone binary system with ternary components. "Ternary" or ("trinary") means that each component can be in one of three states: *up, middle (mid)* and *down*. It turns out that for this system exists a combinatorial invariant by means of which it is possible to count the number $C(r; x)$ of system failure sets with a given number of r components in *up*, x components in *down* and the remaining components in state *mid*. This invariant is called ternary D-spectrum and it is an analogue of signature or D-spectrum for a binary system with binary components. Contrary to D-spectrum, it is not a single set of probabilities, but a collection of such sets. The r-th member of this collection resembles a D-spectrum computed for a special case for which r components are permanently turned into state *up*. If system (network) components are statistically independent and identical, and have probabilities p_2, p_1 and p_0, to be in *up, mid* and *down*, respectively, then the ternary D-spectrum allows obtaining a simple formula for calculating system *DOWN* probability. We consider also so-called ternary importance spectrum by means of which it becomes possible to rank system components by their importance measures. These importance measures are similar to Birnbaum importance measures that are well-known in Reliability Theory. The chapter is concluded by a description of Monte Carlo procedures used for approximating the ternary spectra.

Keywords Ternary components · Ternary network · Signature · Ternary D-spectrum · Failure sets · Ternary importance measure

1.1 Introduction: Networks with Binary and Ternary Components

A standard object in Reliability Theory is a binary monotone system [1]. It consists of n components, and each component is *binary*, i.e. it can be in two states: *up* and *down* denoted 1 and 0, respectively. The state of system components is described by a binary vector $\mathbf{x} = (x_1, x_2, \ldots, x_n)$, where the i-th coordinate $x_i = 1(0)$ if component i is *up* (*down*). Typically, it is assumed that the state of the whole system

I. Gertsbakh et al., *Ternary Networks*, SpringerBriefs in Electrical and Computer Engineering, DOI: 10.1007/978-3-319-06440-6_1,
© The Author(s) 2014

is also binary, i.e. the whole system is operational (*UP*) or nonoperational (*DOWN*). The system state is a binary variable φ having value 1 if the system is *UP* and value 0, if the system is *DOWN*. We assume that system state is completely determined by the state of its components, i.e

$$\varphi = \varphi(\mathbf{x}).$$

Traditionally, Reliability Theory deals with so-called *monotone* systems. These systems have the following natural properties:

(i) $\varphi(0, 0, \ldots, 0) = 0$; system is *DOWN* if all its components are *down*.
(ii) $\varphi(1, 1, \ldots, 1) = 1$; system is *UP* if all its components are *up*.

Let $\mathbf{x}$ and $\mathbf{y}$ be binary state vectors such that $x_i \geq y_i$ and there is one index j for which $x_j > y_j$. Then we say that $\mathbf{x} > \mathbf{y}$.
(iii) If $\mathbf{x} > \mathbf{y}$ than $\varphi(\mathbf{x}) \geq \varphi(\mathbf{y})$.

This means that replacement of an *up* component by a *down* one either does not affect the current system state or can lead to the transition from *UP* to *DOWN*.

It is desirable to have a visual image of a monotone system. In Reliability, examples of monotone systems are always networks. For example, a simple network called *bridge* is a standard illustrative monotone system example in many reliability books. From now on, we will deal only with networks.

In simple words, a network is a collection of nodes (vertices) and edges (links) connecting the nodes. Formally, a network $\mathcal{N}$ is described by means of two sets: $\mathcal{N} = (V, E)$, where V is the node (vertex) set, $|V| = n$, E is the edge (link) set, $|E| = m$.

Physically, nodes serve to denote computers, power stations, individuals, information sources, etc., and edges are simplified images of communication lines, power supply lines, connections of individuals, channels of information exchange, and so on. Traditionally, Reliability Theory considers *binary* networks, i.e. networks whose components have two states: *up* and *down*. For a node v state *down* means that all edges adjacent to v are erased, and v becomes isolated. For an edge $e = (a, b)$ the *down* state means that e does not exist, or is erased, and that there is no direct connection from a to b via this edge. According to the function carried out by the network, we define several performance levels for performance of the whole network. Let us first concentrate on the situation with only two levels denoted as *UP* and *DOWN*. Historically, research efforts have been mainly concentrated around the study of network connectivity in case of unreliable nodes or edges. The definition of network *UP* and *DOWN* states involves a special node set T, $|T| = r$, called *terminal set*. Nodes of this set never fail. The network is *UP* if and only if all nodes in the terminal set are connected to each other by operational edges.

Often network *UP/DOWN* states are associated with the size of its largest (connected) component: network is declared as being *DOWN* if the number of nodes in the largest component becomes less than some critical value L.

Binary networks are strongly simplified *models* of real systems. Probably, the most questionable part of this simplification is the *binary assumption* regarding the state of network components. Indeed, if an edge $e = (a, b)$ represents a water or gas supply pipeline from a to b, in reality this pipeline can be in more than two states: fully operational, 50 or 30 % operational, etc. and completely *down*. An information channel can provide a high-speed transmission rate, transmission with reduced speed, and so on. As it always happens, using "more adequate" model seems on one hand to be more realistic, but on the other hand—involves serious computational difficulties. Take, for example, a simple network—four-node five-edge bridge system. If its edges are binary, then the there are $2^5 = 32$ different component state vectors. If we assume that each edge has three states, we will have $3^5 = 243$ state vectors!

Reliability calculations for binary networks of small-to-medium size (100–200 nodes) has been considerably simplified by using so-called D-spectra techniques, see [4, 5, 8, 9]. Let us describe its main features.

Denote by $e_1, e_2, \ldots, e_n$ network components which are subject to failure. Consider a random permutation of network component numbers:

$$\pi = (e_{i_1}, e_{i_2}, \ldots, e_{i_n}). \tag{1.1}$$

Suppose that all components in (1.1) are *up*, and move along π from left to right, and turn each component from *up* to *down*. Suppose that network state is controlled after each step.

Definition 1.1.1 The ordinal number in the permutation of the component whose turning *down* causes the network to go *DOWN* is called *anchor* and denoted $\delta(\pi)$. #

Assume that all permutations are equally probable and each particular permutation has probability $1/n!$ Then we can speak about the probability $f(x)$ that $\delta(\pi) = x, x = 1, 2, \ldots, n$. Formally, this is the fraction of those permutations which have anchor x. Obviously, the collection $\{f(x), x = 1, \ldots, n\}$ is a proper discrete density.

Definition 1.1.2 The discrete cumulative distribution function $F(x) = \sum_{i=1}^{x} f(i)$ is called the cumulative D-spectrum or simply D-spectrum. ("D" stands for "destruction" of components). #

The D-spectrum is system combinatorial invariant. It does not depend on the probabilistic properties of network components. Now suppose that all network components are stochastically independent and have the same *down* probability q. ($p = 1 - q$ is component *up* probability).

Let us call system binary state vector $\mathbf{y}$ a *failure set* if $\varphi(\mathbf{y}) = 0$.

Denote by $C(x)$ the number of network failure sets which have exactly x components *down* and the remaining $n - x$ components *up*. The probabilistic weight of all failure sets of this type is

$$W(x) = C(x)p^{(n-x)}q^x. \tag{1.2}$$

Then the network is *DOWN* with probability

$$P(DOWN) = \sum_{x=1}^{n} C(x)p^{(n-x)}q^x. \tag{1.3}$$

Fig. 1.1 Information (infection) transmission by a node in *up* (*left*) and in *mid* (*center*)

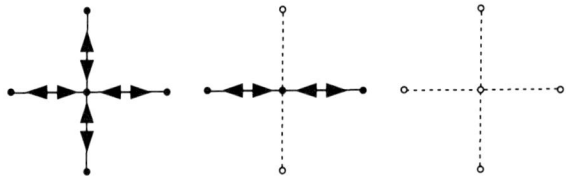

The following combinatorial property of the D-spectrum plays crucial role in all our reliability calculations, see [5–7, 9]:

$$C(x) = F(x)\frac{n!}{x!(n-x)!}. \tag{1.4}$$

Let us note here that the D-spectrum is approximated by means of Monte Carlo simulation. Numerous calculations prove that for a network with node number in the range 80–120, it is possible to obtain in few minutes of CPU-time a rather accurate approximation to the D-spectrum which guarantees about 0.5 % accuracy in estimating network reliability, see [4, 5, 7].

So far we described how the spectrum technique works for binary systems, i.e. for networks with binary components. We mentioned also that "binary world" implies a considerable simplification of reality and may become an inadequate tool for reliability modeling. In the next section we will describe a new collection of reliability models in which we allow the network components to be in three states: *up*, *down* and middle (*mid*), thus introducing so-called ternary (or "trinary") systems. Let us consider several typical examples of ternary networks.

Providing strong and weak connectivity. Suppose we have a network $\mathcal{N} = (V, E)$ with *two* terminal sets T_1 and T_2, $T_1 \subset V$, $T_2 \subseteq V$. We say that the terminals of the set T_1 are strongly connected if all nodes in T_1 are connected to each other by edges in state *up*. We say that nodes in T_2 are weakly connected if all nodes in T_2 are connected to each other by links in state *up* or in *mid*. Now define network states *UP* and *DOWN* as follows: network is *UP* if the set T_1 is strongly connected (event A) *and* the set T_2 is weakly connected (event B). Network is *DOWN* if the event $D = \overline{A \cap B}$ takes place. In practice, strong connection may correspond to fast or more reliable information transmission than in case of "weak" connectivity.

Information/current transmission or percolation on a nonhomogeneous grid. Consider a rectangular two-dimensional grid in which infection is transmitted from infected node to adjacent nodes along edges. Each node (except the border nodes and the corners) is adjacent to four edges. A node is considered *up* if, upon receiving information, it transmits it in all four directions. The node is considered in state *mid* if it transmits information only in horizontal or only in vertical direction. The node which is *down* is isolated and cannot receive/transmit any information, see Fig. 1.1.

Flow networks. In describing flows in networks, edges represent transmission channels, power supply lines, water pipes, etc. In the framework of binary network we are forced to consider each facility $e = (a, b)$ transmitting from node a to node

b in two extreme states: operable and nonoperable, i.e. having some maximal flow capacity $c(e)$ or capacity 0. More adequate is the model in which we introduce for each edge $e = (a, b)$ three levels of performance, say maximal capacity $c_{max}(e)$ (*up* state), intermediate capacity $c_0(e) = 0.5 \cdot c_{max}$ (*mid* state), and zero capacity in case of failure (*down* state).

The price for considering more flexible ternary models is the necessity to introduce more involved combinatorial invariant than D-spectrum. D-spectrum is in fact a single discrete distribution, a vector. For dealing with ternary networks we have to introduce the so-called ternary spectrum, which is a collection of vectors.

1.2 Ternary D-Spectrum

Networks with ternary components

Consider a network with components numbered $1, 2, \ldots, n$. Each component can be in three states: *up, mid* and *down*. j-th component state will be denoted by a ternary variable $y_i = 2, 1$ or 0 if component j is *up, mid* or *down*, respectively. The state of all system components is described by a ternary vector $\mathbf{y} = (y_1, y_2, \ldots, y_n)$.

The state of the ternary system is denoted by φ and depends on component state vector $\mathbf{y}$:

$$\varphi = \varphi(\mathbf{y}).$$

We assume that ternary networks have the same properties (**i**), (**ii**) and (**iii**) postulated for binary systems.

If $\mathbf{y} = (y_1, y_2, \ldots, y_n)$ has the property that $\varphi(\mathbf{y}) = 0$, we call it *failure vector* or *failure set*. If a failure set has r components in *up* (i.e. equal 2), x components in *down* (i.e. equal 0), and remaining components in *mid*, we call it an *(r; x)-failure set*.

Types of permutations Our object of main interest will be random permutations of system component numbers.

First, let us define a *random permutation of r-th type, $r = 0, 1, \ldots, n - 1$*. This is a random permutation of component numbers in which the components on the first r positions are turned into state *up*, and the components on the remaining positions are initially in state *mid*. It has the following form:

$$\pi_r = \{(i_1^{(2)}, i_2^{(2)}, \ldots, i_r^{(2)}; i_{r+1}^{(1)}, i_{r+2}^{(1)}, \ldots, i_n^{(1)}\}.$$

Here the upper index at i_s denotes the state of the component on the s-th position: 2-denotes *up*, 1-denotes *mid*, and 0-denotes *down*. Permutation of 0-type, by the definition, has all its components in state *mid*.

Destruction process Destruction process has several stages denoted by $0, 1, \ldots,$ $n - 1$. Stage r of the destruction process consists of generating random permutation of r-th type, an initial check of system state and sequential destruction of its *mid* components (i.e. turning them from *mid* to *down*) by moving from left to right.

The check of initial system state means checking system state when the r first components in the permutation are *up* and all remaining components are in state *mid*. If for a particular r-permutation the check reveals that the system is already $DOWN$, we say that this permutation has the *anchor* equal zero.

In the process of sequential destruction we are turning components from *mid* to *down* and check the system state after each component destruction. The *anchor* of a permutation of r-th type is the number of components which have been turned *down* when the system was for the first time discovered in state $DOWN$. Let us denote the anchor of r-th type permutation by $\delta(\pi_r)$.

There are two particular types of r-permutations which is convenient to consider separately.

First type is the following. Each permutation of type j, $j = 0, 1, \ldots, r_0$, has the property that the system always is $DOWN$ **before** even a single component in *mid* has been turned *down*. All these permutations have the anchor equal zero. For example, to be in UP state the system must have at least three components *up*, no matter how many components are in state *mid*. Then obviously each permutation of type 0,1 and 2, which has less than than 3 components *up* and all others in *mid* imply that the system is already $DOWN$. Let us call permutations with the above property A-type permutations ("A" stands for "abnormal").

Remark 1.1.1 The maximal r for which A-permutations do exist is denoted by r_0. If there are no permutations of A-type, we put $r_0 = -1$. #

Second type of permutations have the following "survival" property: after turning *down* all $n - r$ components which were in *mid*, the system *remains* in state UP. Call them permutations of type B. Suppose that all r- type permutations for $r > r_{\max}$ are of type B.

Example 1.1.1 Suppose, the system has seven components. Consider a 3-rd type random permutation

$$\pi_3 = (2^{(2)}, 1^{(2)}, 7^{(2)}; 6^{(1)}, 4^{(1)}, 5^{(1)}, 3^{(1)}).$$

In this permutation, the first three positions occupy components 2, 1 and 7 which are in state 2, i.e. *up*. The remaining components are in state 1: component 6 is on the fourth position (from left to right), component 5 is on sixth position, etc. Suppose that we start turning into *down* the components in *mid* by moving from left to right. Suppose that after doing this to components 6 and 4, the system remained in UP. When we turned *down* component 5, the system went $DOWN$. Then the anchor of this permutation $\delta(\pi_3) = 3$. #

From now on, we assume that we deal only with "regular" permutations which are not of type A and not of type B, i.e. we consider only r-permutations for $r_0 < r \leq r_{\max}$.

Assume that on each stage of the destruction process, all r—permutations are equally probable and each particular permutation appears with probability $1/n!$. Let $f_r(y)$ be the probability that the anchor of r-th type permutation equals y, $y = 0, 1, 2, \ldots, n - r$:

$$f_r(y) = P(\text{in } \pi_r, \text{ the anchor } \delta(\pi_r) = y).$$

Since the total number of permutations on for each r is $n!$, for $y = 0, 1, \ldots, n - r$

$$f_r(y) = \frac{\text{The number of } r\text{-permutations with } \delta(\pi_r) = y}{n!}. \qquad (1.5)$$

Remark 1.2.2 All $n!\, r$-permutations are classified during the destruction process into $n - r + 2$ classes: $n - r + 1$ classes—according to the anchor position, and one extra class—for r-permutations where the system remains *UP* if all $n - r$ *mid* components have been turned *down*, i.e the anchor was not found. Some of these classes may be empty. #

Definition 1.2.3 Let $r_0 < r \le r_{max}$. Then let

$$F_r(x) = \sum_{y=0}^{x} f_r(y), x = 0, 1, 2, \ldots, n - r.$$

is called cumulative r-spectrum. Obviously, $F_r(0) = f_r(0)$, and $F_r(x) \le 1$.

Definition 1.2.4 The collection of all r-spectra $\mathcal{T}sp = \{F_r(x)\}$ for $r_0 < r \le r_{max}$, is called *ternary D-spectrum*. In words: ternary D-spectrum is the collection of all cumulative r-spectra. #

Denote by $C(r; x)$ the number of network failure sets with r components *up* and x components *down* (and therefore $(n - x - r)$ components in *mid*).

Theorem 1.2.1 For $r_0 < r \le r_{max}$ and $x = 0, 1, 2, \ldots, n - r$,

$$C(r; x) = F_r(x) \cdot \frac{n!}{r!x!(n - x)!}. \qquad (1.6)$$

Proof Let us count all r-type permutations having anchor j, $0 \le j < x$. After the destruction is completed, there will be $n! f_r(j)$ such permutations. Each of them has the following form:

$$(i_1^{(2)}, i_2^{(2)}, \ldots, i_r^{(2)}; i_{r+1}^{(0)}, \ldots, i_{r+j}^{(0)}; i_{r+j+1}^{(1)}, \ldots, i_n^{(1)}) \qquad (*).$$

Now let us turn *down* the components which are on the positions $r + j + 1, \cdots,$ $r + j + 2, r + j + x$. Then each of permutations (*) takes the form

$$(i_1^{(2)}, i_2^{(2)}, \ldots, i_r^{(2)}; i_{r+1}^{(0)}, \ldots, i_{r+x}^{(0)}; i_{r+x+1}^{(1)}, \ldots, i_n^{(1)}), \qquad (**)$$

i.e. the first r positions occupy components in *up*, the next x positions—components in *down*, and the remaining—in *mid*. Obviously, all these permutations are $(r; x)$-failure sets. Therefore, all r-permutations produce

$$n![f_r(0) + f_r(1) + \cdots + f_r(x)] = n!F_r(x) \qquad (1.7)$$

$(r; x)$-failure sets. But each particular failure set will be repeated $r!x!(n - r - x)!$ times. Therefore, the number of original $(r; x)$-failure sets produced in the above process equals

$$F_r(x) \cdot \frac{n!}{r!x!(n - r - x)!} = C_1 \qquad (1.8)$$

and therefore $C(r; x) \geq C_1$. On the other hand, each $(r; x)$-failure set has its "representatives" in permutations (**). Therefore, $C(r; x) = C_1$. #

Example 1.2.2 Let us consider a parallel system of three identical and independent components, a, b and c. Each component can be in three states: *up*, *down* and *mid*. In state *up* each component has capacity 2, in state *mid*—capacity 1 and in state *down* capacity 0. The total flow provided by the system ranges therefore from 6 to 0. System state is defined as a function of component capacities, as follows. If the sum of all component capacities is less or equal 2, the system is *DOWN*, otherwise, it is *UP*. Note that if all components are in *mid* state, the system is *UP*. (There are no A-type permutations). Let us start the destruction process. Set $r = 0$, i.e. all components are in *mid*. After turning *down* the first component, the system always gets *DOWN*. Therefore, $f_0(1) = 1$, $f_0(2) = f_0(3) = 0$ and $F_0(1) = F_0(2) = F_0(3) = 1$. Now, by the theorem above,

$$C(r = 0; x = 1) = F_0(1) \cdot \frac{3!}{0!1!2!} = 3,$$

and the corresponding failure sets are $(a = 0, b = 1, c = 1)$, $(a = 1, b = 0, c = 1)$, $(a = 1, b = 1, c = 0)$. Similarly, $C(r = 0, x = 2) = 3$, and the corresponding failure sets are $(a = 0, b = 0, c = 1)$, $(a = 0, b = 0, c = 1)$, $(a = 1, b = 0, c = 0)$. Finally, $C(r = 0, x = 3) = 1$ which corresponds to the unique failure set $(a = b = c = 0)$ with all three components *down*.

Let us consider the next step in the destruction process with $r = 1$ components *up* and two in *mid*. Such a system fails only after exactly two components from *mid* are turned to *down*. We see therefore that here $f_1(1) = 0$, $f_1(2) = 1$ and therefore $F_1(1) = 0$, $F_1(2) = 1$. From here we find out that $C(r = 1; x = 2) = 3$. There are exactly three *DOWN* sets with 1 component *up* and two *down*: $(a = 2, b = c = 0)$, $(a = c = 0, b = 2)$, $(a = b = 0, c = 2)$. So, we found out all $7 + 3 = 10$ failure sets of the system. #

Theorem 1.2.1 opens the way to compute system DOWN probability if the system consists of n identical and statistically independent components, and a component is in state up with probability p_2, in down—with probability p_0 and in mid—with probability $p_1 = 1 - p_0 - p_2$.

Obviously, the probabilistic weight of a particular $(r; x)$-failure set equals

$$W(C(r; x)) = p_2^r \cdot p_0^x \cdot p_1^{(n-r-x)}. \qquad (1.9)$$

Remember that so far we did not consider failure sets corresponding to permutations of type A and B. Permutations of type B do not produce failure sets. Each j-type permutation of type A produces $n!/(j!(n - j - x)!x!)$ failure sets with j components *up*, $(n - j - x)$ in *mid*, and x components *down*. Here x is running from 0 to $n - j$. The total probabilistic weight of all failure sets produced by A-type permutations equals

$$W_A = \sum_{j=0}^{r_0} \sum_{x=0}^{n-j} p_2^j p_1^{(n-j-x)} p_0^x \cdot \frac{n!}{j!x!(n - j - x)!} \qquad (1.10)$$

Now note that the network is *DOWN* iff it is in one of its failure sets. Put $F_r(x) \equiv 1$ for $r \le r_0, \forall x$. Combining all together, we arrive at

Theorem 1.2.2

$$P(DOWN) = \sum_{j=0}^{r_{max}} \sum_{x=0}^{n-j} F_j(x) \cdot p_2^j \cdot p_0^x \cdot p_1^{(n-j-x)} \frac{n!}{j!x!(n - j - x)!} \# \qquad (1.11)$$

Example 1.2.3 $s - t$ flow in a three component network.
The network has three edges: $a = (s, v), b = (v, t)$ and $c = (s, t)$, see Fig. 1.3. Each edge has three states: capacity 2 (*up*), capacity 1 (*mid*) and capacity 0 (*down*). The maximal $s - t$ flow equals 4. Network is *DOWN* if the $s - t$ flow is 0, 1 or 2. Let us consider all permutations of type $r = 1$, i.e. having one (first) component *up* and two others in *mid*. There are $3! = 6$ such permutations. Four of them have anchor zero:

$$(a = up, b = mid, c = mid), (a = up, c = mid, b = mid),$$

$$(b = up, a = mid, c = mid), (b = up, c = mid, a = mid).$$

Therefore, $f_1(0) = 4/6 = \text{definition} = F_1(0)$. Therefore, by (1.6),

$$C(1; 0) = \frac{4}{6} \cdot \frac{3!}{0!1!2!} = 2.$$

Indeed, there are only two failure sets with exactly one component *up* and two in *mid*:

$$(a = up, b = mid, c = mid), (a = mid, b = up, c = mid).\#$$

Fig. 1.2 $s - t$ flow
in a triangular network

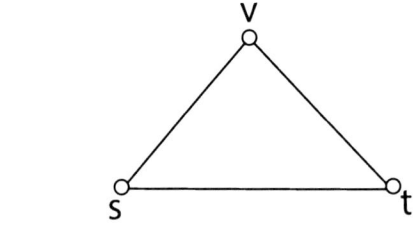

Fig. 1.3 Bridge network. The
up edges are shown *bold*,
down edges are *dashed*. *Left
upper figure* shows bridge in
UP, three others in *DOWN*

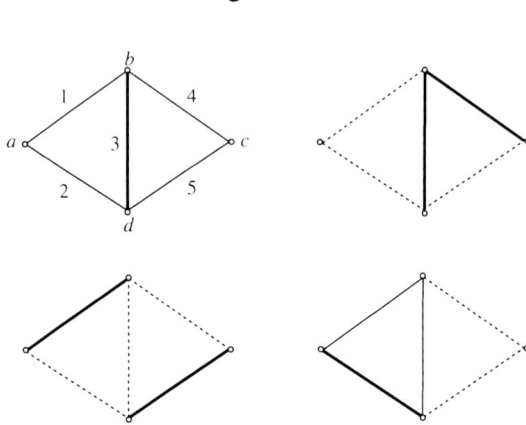

Example 1.2.4 Bridge network.

Let us consider a bridge network shown on Fig. 1.3.

Components subject to failure are the edges numbered 1, 2,..., 5. Each edge has three states: *up, mid* and *down*. The network is, by definition, *UP* iff nodes b and d are connected by edges in *up* **and** all nodes are connected to each other. The ternary spectrum is presented in Table 1.1.

For $r = 0, x = 0$, all components are in *mid*, and there is no strong connection between b and d. Therefore, the bridge is *DOWN*. This explains the $r = 0$ row in Table 1.1. All 0-type permutations are A-type permutations.

When $r = 1$ and $x = 0$ (one edge is *up*, all other edges are in *mid*), the bridge can be *UP* only if the *up* edge is on the place of edge 3, which happens with probability $1/5 = 0.2$. This explains the first entry in line 2. Checking all permutations we see that turning one edge from *mid* to *down* ($x = 1$) does not change the *DOWN* probability.

Now consider $r = 1$ and $x = 2$—one edge *up* and two—*down*. It is easy to check that 1/3 of all permutations providing *UP* on the previous step (i.e. for $x = 1$), now turn into *DOWN*, and thus $P(DOWN) = 96/120 + 8/120 = 104/120$.

Analyzing the remaining r-permutations along similar lines, we find *DOWN* probabilities for $r = 2$ and $r = 3$ rows. When $r = 4$, i.e. 4 components are *up*, the bridge is always *UP* and $P(DOWN) = 0$. Therefore, in this example $r_{max} = 3$. #

Table 1.1 Ternary D-spectrum $F_r(x)$ of bridge network

r	$x = 0$	$x = 1$	$x = 2$	$x = 3$	$x = 4$	$x = 5$
$r = 0$	1	1	1	1	1	1
$r = 1$	0.8	0.8	$0.867 = 104/120$	1	1	–
$r = 2$	0.4	0.4	0.6	1	–	–
$r = 3$	0	0	0.2	–	–	–

1.3 Dynamic Version of Theorem 1.2.2

Formula (1.11) can be rewritten in an equivalent "dynamic" form to include the time factor. Suppose we have n statistically independent and stochastically identical point processes $\{\chi_i(t), t > 0, i = 1, \ldots, n\}$. Each $\chi_i(t)$ is a decreasing, time continuous process with three states: 2, 1 and 0. State 0 is absorbing. Each trajectory of $\chi_i(t)$ starts at $t = 0$ in state 2, jumps into state 1 and later gets absorbed in state 0. At any time instant $t > 0$, $\chi_i(t)$ is in one of its three states with probabilities $p_2(t)$, $p_1(t)$ and $p_0(t)$, respectively. Obviously

$$p_0(t) + p_1(t) + p_2(t) = 1, t \geq 0.$$

Let $\tau_2(i)$ be the sojourn time of $\chi_i(t)$ in state 2. Then $P(\tau_2 \leq t) = 1 - p_2(t)$. Let $\tau_1(i)$ be the sojourn time of $\chi_i(t)$ in state 1. Then the event $\{\tau_2(i) + \tau_1(i) \leq t\}$ means that at time $t + 0$, $\chi_i(t)$ has already left state 1, and is therefore in state 0, i.e.

$$P(\tau_2(i) + \tau_1(i) \leq t) = p_0(t).$$

Note that if at time t the system is *DOWN*, then its failure-free operation time τ_{up} does not exceed t.

Therefore, for any $t > 0$,

$$P(\tau_{up} \leq t) =$$

$$\sum_{j=0}^{r_{max}} \sum_{x=0}^{n-j} F_j(x) \frac{n!}{j! x! (n - j - x)!} \cdot [p_2(t)]^j \cdot [p_0(t)]^x \cdot [1 - p_0(t) - p_2(t)]^{(n-j-x)}.$$

(1.12)

Example 1.3.1 Dynamics of strong and weak connection in a triangular network.
Consider a triangular network shown on Fig. 1.3. It has three nodes s, v, t and three edges: $e_1 = (s, v)$, $e_2 = (v, t)$ and $e_3 = (s, t)$. Each edge has three states: *up*, *mid* and *down*, denoted as 2, 1 and 0, respectively. The state of all edges will be denoted by a three digit ternary number. For example, $\mathbf{x} = (012)$ means that edge e_1 is *down*, edge e_2 is *mid* and edge e_3 is *up*.

By definition, the network is *UP* if there is a strong connection of s and t, *and* all nodes are connected. Strong connection of s and t means that there is a $s - t$ path of edges in *up* only. In our case it means that either e_3 is *up*, or e_1 and e_2 are *up*, or both. *DOWN* means that either there is no strong $s - t$ connection, or one node is isolated, or both.

The total number of states is $3^3 = 27$. It is easy to check that there are the following 17 system states for *DOWN*:

(000), (001), (002), (010), (011), (020), (021), (100), (101), (110), (111),
(120), (121), (200), (201), (210), (211).

It is easy to count directly from this list the numbers $C(r; x)$ of failure sets having r components *up* and x *down*. We have

$$C(0; 3) = 1, C(0; 2) = 3, C(0; 1) = 3, C(0; 0) = 1, C(1; 2) = 3,$$
$$C(1, 1) = 4, C(1; 0) = 2.$$

For example, states (021), (120), (210) and (201) all are of type $C(1; 1)$.

Now assume that all components are identical and independent, and the behavior of each one is described by a three state random process $\chi(t)$. It spends time τ_2 in state 2, τ_2 has Gamma distribution with parameters $\lambda = 1, k_2 = 2$ and therefore (see [1]), $P(\tau_2 > t) = exp(-t)(1 + t)$. From state 2 (*up*), $\chi(t)$ jumps into state 1 (*mid*) and spends in it some random time τ_1, independent on τ_2, which has Gamma distribution with parameters $\lambda = 1, k_1 = 3$. Then it follows that $\tau_1 + \tau_2$ is Gamma distributed with parameters $\lambda = 1, k = k_1 + k_2 = 5$.

Obviously, at time t a component is in state 2 if $\tau_2 > t$, and is in state 0 if $\tau_2 + \tau_1 \leq t$. From here it follows that

$$p_2(t) = e^{-t}(1 + t); p_0(t) = 1 - e^{-t}(1 + t + t^2/2! + t^3/3! + t^4/4!);$$
$$p_1(t) = 1 - p_0(t) - p_2(t).$$

Collecting all together we obtain from (1.12) the expression for system *DOWN* probability at time t:

$$P(DOWN; t) = p_0^3(t) + 3p_0^2(t)p_1(t) + 3p_0(t)p_1^2(t) + p_1^3(t) + 3p_2(t)p_0^2(t)$$
$$+ 4p_2(t)p_1(t)p_0(t) + 2p_2(t)p_1^2(t).$$

Figure 1.4 shows the graph of system reliability $R(t) = 1 - P(DOWN; t)$. #

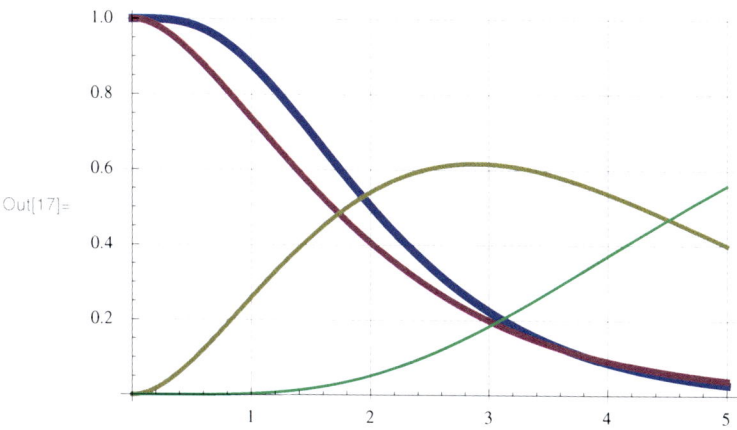

Fig. 1.4 *Thick* curve is $R(t) = 1 - P(DOWN; t)$. *Thin* curves are the state probabilities $p_2(t), p_1(t), p_0(t)$

1.4 *P(UP)* and Path Sets

A vector of component states $\mathbf{y} = (y_1, y_2, \ldots, y_n)$ is called a *path* or a *path set* if $\varphi(\mathbf{y}) = 1$. Let $\mathbf{y}_{(j;x)}$ be a path vector which has j components *up*, x components *down*, and the remaining $(n - x - j)$ components in *mid*. Denote by $V(j; x)$ the number of path sets of type $\mathbf{y}_{(j;x)}$, $r_0 < j \leq r_{max}$. For $r > r_{max}$, put $F_r(x) \equiv 0$. Note that for $j \leq r_0$, we have set $F_j(x) \equiv 1$.

Theorem 1.2.3

(i)

$$P(UP) = \sum_{j=0}^{n} \sum_{x=0}^{n-j} \frac{n!}{j!x!(n-j-x)!} \cdot (1 - F_j(x)) p_2^j p_0^x p_1^{(n-j-x)}. \qquad (1.13)$$

(ii)
For $0 \leq j \leq n$,

$$V(j; x) = \frac{n!}{j!x!(n-j-x)!} \cdot (1 - F_j(x)). \qquad (1.14)$$

Proof

(i)

$$P(UP) = 1 - P(DOWN) = (p_2 + p_1 + p_0)^n - P(DOWN)$$

$$= \sum_{j=0}^{n} \sum_{x=0}^{n-j} \frac{n!}{j!x!(n-j-x)!} p_2^j p_1^x p_0^{(n-j-x)}$$

$$-\sum_{j=0}^{n}\sum_{x=0}^{n-j}F_j(x)\frac{n!}{j!x!(n-j-x)!}\cdot p_2^j\cdot p_0^x\cdot p_1^{(n-j-x)}.$$

Simplifying this expression we obtain (1.13).

(ii) The network is *UP* iff it is in one of its path sets. Therefore, the coefficients at $p_2^j p_1^{n-j-x} p_0^x$ should be the number of the path sets with j components *up*, x components *down* and the remaining $n - j - x$ components in *mid*. This proves (1.14). #

1.4.1 Remarks

1. The ternary D-spectrum is system *combinatorial invariant*. It depends only on its structure and failure definition. No information about component reliability is needed for calculating the ternary spectrum. Contrary to the D-spectra for binary systems considered in literature (also known as *signature* or *internal distribution*, see [3–5, 10, 11]), the ternary spectrum is not a single discrete cumulative distribution function (CDF) but a *collection* of such CDF's, some of which are not proper, i.e. some of the CDF's are strictly less than 1. The cumulative D-spectrum of a binary system is a vector of dimension n, while the ternary spectrum is a set of k vectors, $k < n$, and all of them, except the first, may have dimension strictly less than n.

2. The complexity of ternary spectrum vs regular binary spectrum can be explained by the fact that ternary system is considerably more complex than the similar binary system with the same number of components. For example, hypercube of order 4 with ternary edges, which will be considered in the next section, has 32 edges and total number of elementary states 3^{32}. This is greater than the number of states of the same hypercube with binary edges by factor of about 430,000.

3. Signature or D-spectrum of a system with binary components has an important property first discovered by Samaniego [10]. If component lifetimes X_i are i.i.d. with CDF $F(t)$, the CDF $G(t)$ of system life-time τ can be expressed as a convex combination of signature coefficients $s_1, s_2, \ldots, s_n$ and respective order statistics:

$$P(\tau > t) = \overline{G(t)} = \sum_{i=1}^{n} s_i \sum_{j=0}^{i-1}[F(t)]^j[1 - F(t)]^{(n-j)}\frac{n!}{j!(n-j)!}, \qquad (*)$$

see [11], page 26. Similar property does not exist for ternary systems. A certain resemblance to it can be seen in formula (1.12).

A very important fact is that the ternary spectrum, similarly to binary spectrum or signature, allows to count the number of failure sets and path sets in the system with given number of *up* and *down* components, see Sects. 1.2 and 1.3. For example, consider a binary monotone system of n components. Let its signature be

$\mathbf{s} = (s_1, s_2, \ldots, s_n)$. Denote by $S(x)$ the cumulative signature $S(x) = \sum_{i=1}^{x} s_i$. It is identical to cumulative D-spectrum. Let $C(x)$ be the number of systems failure sets with x components *down* (and the remaining $(n - x)$ in state *up*). The following equality is very useful in reliability calculations:

$$C(x) = S(x) \frac{n!}{x!(n - x)!},$$

compare with [11], page 80. One can derive it directly from the above formula (*) by changing the summation order.

This formula is very similar to our formula (1.6):

$$C(r; x) = F_r(x) \frac{n!}{r!x!(n - r - x)!}$$

4. Let us have another look at the formulas for $C(x)$ and $C(r; x)$. The first can be rewritten as

$$S(x) = \frac{C(x)}{\frac{n!}{x!(n-x)!}}$$

The denominator is the total number of ways to choose randomly and equiprobable an x-subset from a set of n items. In our situation, this is the number of ways to "nominate" x *down* components in a set of n components. Some of these x-subsets are failure sets, and $S(x)$ is the fraction of these failure sets. A nontrivial fact is that $S(x) = s_1 + s_2 + \cdots + s_x$.

Similarly, formula (1.6) can be rewritten as

$$F_r(x) = \frac{C(r; x)}{\frac{n!}{r!x!(n-r-x)!}},$$

which says that the fraction of (r, x)-failure sets among all random and equiprobable chosen (r, x)-subsets from a set of n items equals $F_r(x) = f_r(0) + f_r(1) + \cdots + f_r(x)$.

1.5 Component Importance

1.5.1 Introduction

In this section we present a very useful reliability characteristic—component importance. We will assume that the system (the network) consists of independent components. In binary case, when each component has two states *up* and *down*, the most known is the so-called Birnbaum Importance Measure (BIM). For component m, it is defined (see [1], Chap. 1, and [2]) as

$$BIM_m = \frac{\partial R(p_1, \ldots, p_n)}{\partial p_m}, \tag{1.15}$$

where $R(p_1, \ldots, p_n)$ is network reliability, i.e. its *UP* probability as a function of component reliabilities $p_i, i = 1, \ldots, n$.

Using pivotal decomposition, it is easy to obtain that

$$BIM_m = R(p_1, \ldots, 1_m, \ldots, p_n) - R(p_1, \ldots, 0_m, \ldots, p_n)$$
$$= G(p_1, \ldots, 0_m, \ldots, p_n) - G(p_1, \ldots, 1_m, \ldots, p_n), \tag{1.16}$$

where $R(p_1, \ldots, 1_m, \ldots, p_n)$ is the reliability of the network with the component m being *up*, and $R(p_1, \ldots, 0_m, \ldots, p_n)$ is the reliability of the network with the component m being *down*. For convenience we represented $R(\cdot)$ as $1 - G(\cdot)$ where $G(\cdot)$ is network *DOWN* probability.

BIM_m has transparent probabilistic meaning: it is the gain in network reliability received from replacing a *down* component m by an absolutely reliable one. BIM_m, being partial derivative, gives an approximation to the system reliability increment δR resulting from reliability increment of component m by δp_m. This increment equals $\delta R(\cdot) \approx BIM_m \cdot \delta p_m$.

In binary situation, increment of component *up* probability p by Δ means necessarily *decrease* by Δ of the *down* probability $q = 1 - p$. In ternary case there is a complication: suppose we increase p_2 by Δ. Then we have to reduce p_1 together with p_0 by the same quantity, and this can be done in several ways. For example, we can reduce p_0 by Δ and leave unchanged p_1. Or we can decrease p_1 by Δ and leave p_0 unchanged, or we can reduce both p_1 and p_0 by $\Delta/2$, and so on.

Before we give the definition of component important measure it is necessary to adjust network reliability expression to ternary case. For ternary networks, we assume that all components subject to failure (nodes or edges) are independent and *identical*, which means that each component has its state distribution determined by a three-component vector

$$\mathbf{p}^i \equiv \mathbf{p} = (p_2, p_1, p_0 = 1 - p_2 - p_1), i = 1, \ldots, n,$$

where $p_2 = P(up)$, $p_1 = P(mid)$, and $p_0 = P(down)$, Accordingly, instead of writing $P(UP) = R(p_1, p_2, \ldots, p_n)$ and $P(DOWN) = G(p_1, p_2, \ldots, p_n)$ for binary network, we will use for ternary network the notation

$$P(DOWN) = G(\mathbf{p}, \mathbf{p}, \ldots, \mathbf{p}).$$

Let us introduce several importance measures for system components reflecting the reliability increase rate for *down-up* and *down-mid* replacements.

Definition 1.5.1-A

The *up–down* ternary importance measure (TIM) of component m is defined as

$$TIM_m(up–down) = G(\mathbf{p}, \ldots, 0_m, \ldots, \mathbf{p}) - G(\mathbf{p}, \ldots, 2_m, \ldots, \mathbf{p}). \tag{1.17}$$

Here the first term in (1.17) is network *DOWN* probability given that the m-th component is turned permanently *down*, and the second term is network *DOWN* probability given that the m-th component is permanently in state *up*. #

Definition 1.5.1-B

The *mid–down* ternary importance measure (TIM) of component m is defined as

$$TIM_m(mid\text{–}down) = G(\mathbf{p}, \ldots, 0_m, \ldots, \mathbf{p}) - G(\mathbf{p}, \ldots, 1_m, \ldots, \mathbf{p}). \quad (1.18)$$

Here the first term in (1.18) is network *DOWN* probability given that the m-th component is turned permanently *down*, and the second term is network *DOWN* probability given that the m-th component is permanently in state *mid*. #

Definition 1.5.1-C

The *up–mid* ternary importance measure (TIM) of component m is defined as

$$TIM_m(up\text{–}mid) = G(\mathbf{p}, \ldots, 1_m, \ldots, \mathbf{p}) - G(\mathbf{p}, \ldots, 2_m, \ldots, \mathbf{p}). \quad (1.19)$$

Here the first term in (1.18) is network *DOWN* probability given that the m-th component is turned permanently *mid*, and the second term is network *DOWN* probability given that the m-th component is permanently in state *up*.

Obviously,

$$TIM_m(up\text{–}mid) = TIM_m(up\text{–}down) - TIM_m(mid\text{–}down).\#$$

The problem in calculating the component importance is that usually the reliability function $G(\cdot)$ is not available in closed analytic form. The *TIM-spectrum*, which we will define below allows to evaluate TIM's by means of Monte Carlo simulation, will be based on introducing another combinatorial invariant—the so called TIM-spectrum. Its construction resembles the construction of binary importance spectrum described in [5, 6].

1.5.2 TIM-Spectrum

Our purpose is to develop a combinatorial measure of component importance in absence of an analytic expression for system reliability as a function of component failure probabilities. For this purpose let us remind the notion of the r-permutation defined in Sect. 1.2, $r_0 < r \leq r_{\max}$. *In the process of component destruction*, this permutation has on its first r positions the numbers of components in state *up*, on the next x positions are component numbers which are in state *down*, and on the remaining $(n - r - x)$ positions are numbers of components which are in state *mid*. Before the destruction starts, there are no components in *down*.

Definition 1.5.2-a

(a) Let $f_r(y|m - down)$ be the probability that the anchor of r-th type permutation equals y, given component m is *down*.

(b) Let $f_r(y|m - mid)$ be the the the probability that the anchor of r-th type permutation equals y, given component m is *mid*.

(c) Let $f_r(y|m - up)$ be the probability that the anchor of r-th type permutation equals y, given component m is *up*. #

Note that $f_r(y|m - state) \cdot n!$ gives the number of r-permutations with the anchor $\delta(\pi_r) = y$, and the component m being in state *down*, *mid*, or *up*, respectively.

Definition 1.5.2-b

Let $r_0 < r \leq r_{max}$. Then for $x = 0, 1, 2, \ldots, n - r$; $m = 1, 2, \ldots, n$:

$$(a)\ F_r(x|m - down) = \sum_{y=0}^{x} f_r(y|m - down),$$

is called the *TIM-down-spectrum* of the system;

$$(b)\ F_r(x|m - up) = \sum_{y=0}^{x} f_r(y|m - up),$$

is called the *TIM-up-spectrum* of the system;

$$(c)\ F_r(x|m - mid) = \sum_{y=0}^{x} (f_r(y|m - mid)) \cdot \frac{n - r - y}{n - r},$$

is called the *TIM-mid-spectrum* of the system. #

Note that (**a**) is the probability that if in an r-permutation x components are *down* and the component m is also *down*, then the system is *DOWN*. Similarly, (**b**) is the probability that if in an r-permutation x components are *down* and the component m is *up*, then the system is *DOWN*.

Let us explain (**c**), which differs from (**a**) and (**b**). When component m is *down* (*up*), then during the destruction process, it remains *down* (*up*). But if m is in *mid*, then during the destruction process, it turns into the *down*. Let us consider now an r-permutation and start the destruction process, i.e. turn the first *mid* component into *down*. Clearly, component m may be on each of $n - r$ places with equal probability $\frac{1}{n-r}$. After the first component is turned into *down*, component m remains in *mid* with probability $\frac{n-r-1}{n-r}$. When the second *mid* component is turned to *down*, m remains *mid* with probability $\frac{n-r-2}{n-r}$, and so on. From this it follows, that $F_r(x|m - mid)$ expresses the probability that if in an r-permutation x components are *down* and the component m is in *mid*, then the system is *DOWN*.

Theorem 1.5.1 Denote by $C(r, x; m - up)$ the number of failure sets with r components up and x components $down$ (and therefore $n - x - r$ components in mid), in which component m is among components in up. In similar way we define $C(r, x; m - down)$ and $C(r, x; m - mid)$.

Then

(i) $C(r, x; m - up) = F_r(x|m - up) \cdot \frac{n!}{r!x!(n-r-x)!}$

(ii) $C(r, x; m - down) = F_r(x|m - down) \cdot \frac{n!}{r!x!(n-r-x)!}$

(iii) $C(r, x; m - mid) = F_r(x|m - mid) \cdot \frac{n!}{r!x!(n-r-x)!}$

Proof

The meaning of $F_r(x|m - up)$ (or $F_r(x|m - down)$, $F_r(x|m - mid)$) is similar to $F_r(x)$ with the only difference that it is given that m is up (or $down$, mid, respectively). That does not affect the proof of the Theorem 1.2.1. #

Now we can turn to computing the probabilities $G(\mathbf{p}, \ldots, 0_m, \ldots, \mathbf{p})$, $G(\mathbf{p}, \ldots, 1_m, \ldots, \mathbf{p})$, and $G(\mathbf{p}, \ldots, 2_m, \ldots, \mathbf{p})$ for the TIM's (1.17–1.19). Note that the probabilistic weight of a particular failure set $C(r, x; m - up)$ given component m is up equals

$$W(C(r, x; m - up)) = p_2^{r-1} \cdot p_0^x \cdot p_1^{n-r-x}.$$

Similarly,

$$W(C(r, x; m - down)) = p_2^r \cdot p_0^{x-1} \cdot p_1^{n-r-x},$$

and

$$W(C(r, x; m - mid)) = p_2^r \cdot p_0^x \cdot p_1^{n-r-x-1}.$$

Let us remind that each j-type permutation of type A, with j components up, x components $down$ and $(n - j - x)$ components in mid produces $\frac{n!}{j!(n-j-x)!x!}$ failure sets.

The probabilistic weight of a particular failure set given component m is $down$, obviously, is the same as in the common case above.

Then combining all together and setting for $r \leq r_0$

$$F_r(x|m - down) \equiv 1, \quad F_r(x|m - up) \equiv 1, \quad , F_r(x|m - mid) \equiv 1$$

we arrive at

Theorem 1.5.2

(i) $G(\mathbf{p}, \ldots, 0_m, \ldots, \mathbf{p}) = \sum_{j=0}^{r_{max}} \sum_{x=1}^{n-j} \frac{F_j(x|m-down) \cdot n!}{j!x!(n-j-x)!} p_2^j p_0^{x-1} p_1^{(n-j-x)}$;

(ii) $G(\mathbf{p}, \ldots, 1_m, \ldots, \mathbf{p}) = \sum_{j=0}^{r_{max}} \sum_{x=1}^{n-j} \frac{F_j(x|m-mid) \cdot n!}{j!x!(n-j-x)!} p_2^j p_0^x p_1^{(n-j-x-1)}$;

(iii) $G(\mathbf{p}, \ldots, 2_m, \ldots, \mathbf{p}) = \sum_{j=0}^{r_{max}} \sum_{x=1}^{n-j} \frac{F_j(x|m-up) \cdot n!}{j!x!(n-j-x)!} p_2^{j-1} p_0^x p_1^{(n-j-x)}$. #

Example 1.5.1 Let us consider a small flow network with three edges a, b and c. Edges $a = (s, v)$ and $b = (v, t)$ are in series, and are parallel to edge $c = (s, t)$, see Fig. 1.3. Each edge can be in three states: *up, mid* and *down* with respective flow capacities 2, 1 and 0. System is *UP* if the maximal $s - t$ flow is at least 2. Otherwise the system is *DOWN*. Below is the list of all 13 network *DOWN* states (failure vectors):

$$(0, 0, 0), (0, 0, 1), (0, 1, 0), (0, 2, 0), (1, 0, 0), (2, 0, 0), (0, 1, 1), (0, 2, 1),$$
$$(1, 0, 1), (2, 0, 1), (1, 1, 0), (1, 2, 0), (2, 1, 0).$$

(Here on the first position is the state of component a, on the second—the state of component b, and on the third—of component c.)

Let us calculate the *up–down* TIMs of all edges using the Definition 1.5.1-A and the above list of failure vectors.

By symmetry. $TIM_a = TIM_b$, and

$$TIM_a = (p_0^2 + 2p_0 p_1 + p_1^2 + p_0 p_2 + p_1 p_2) - (p_0^2 + 2p_0 p_1)$$
$$= p_1^2 + p_0 p_2 + p_1 p_2.$$
$$TIM_c = p_2^2 + 2p_0 p_1 + 2p_0 p_2 + p_1^2 + 2p_1 p_2.$$

We see that

$$TIM_a = TIM_b < TIM_c,$$

which is in accord with our intuition.

Let us now demonstrate the calculation of TIM_a using (**i**) and (**ii**) of Theorem 1.5.2.

In total, there are the following six permutations of all components:

$$per1 = (a, b, c), per2 = (a, c, b), per3 = (b, a, c), per4 = (b, c, a),$$
$$per5 = (c, a, b), per6 = (c, b, a).$$

Suppose that a is *down*. Put $r = 0$. Let us count the number of r-permutations $F_r(x|a - down) \cdot 3!$ having x components *down* **and** a *down*. We have $F_0(1|a - down)3! = 2$—for *per1* and *per2*.

$F_0(2|a - down)3! = 4$—for *per1, per2, per3, per5*.

$F_0(3|a - down)3! = 6$—for all permutations.

Now let $r = 1$. $F_1(1|a - down)3! = 1$ - for *per1*. $F_1(2|a - down)3! = 2$—for *per3, per4*. For $r = 2$ all permutations produce *UP*.

$$G(a - down; \mathbf{p}) = \frac{2}{0!1!2!} p_1^2 + \frac{4}{0!2!1!} p_0 p_1 + \frac{6}{0!3!0!} p_0^2 + \frac{1}{1!1!1!} p_2 p_1 + \frac{2}{1!2!} p_2 p_0.$$

Suppose now that a is *up*. Let us count the number of r-permutations $F_r(x|a -$
$up) \cdot 3!$ having x components *down* **and** a—up. The case $r = 0$ is not relevant. Let
$r = 1$. Then
  $F_1(1|a - up)3! = 2$—for *per1, per2*. $F_1(2|a - up)3! = 2$—for *per1, per2*. For
$r = 2$, all r-permutations produce *UP*.
  Therefore

$$G(a - up; \mathbf{p},\mathbf{p}) = \frac{2}{1!1!1!} p_0 p_1 + \frac{2}{1!2!} p_0^2.\#$$

Finally,

$$TIM_a(up\text{–}down) = G(a - down; \mathbf{p},\mathbf{p}) - G(a - up; \mathbf{p},\mathbf{p}) = p_1^2 + p_1 p_2 + p_0 p_2.\#$$

For systems having more than three components, manual calculations of the *TIMs*
becomes too involved. There is however an efficient way of numerical estimation
of ternary importance spectrum by means of Monte Carlo simulation. In the next
section we will show how the importance spectrum can be approximated by means
of a small modification of the principal Monte Carlo algorithm for approximating
the ternary spectrum.

1.6 Ternary Spectrum Monte Carlo

Let us remind that there are two particular types of r-permutations. We call them A-
type and B-type permutations. A-type has the following property: for $r = 0, 1, \ldots, r_0$
the system always is *DOWN* before even a single component in *mid* has been turned
down. In other words, all permutations have the anchor equal zero.
  In B-type permutation, after turning all *mid* components into *down*, the system
remains in *UP* state. Suppose that all r-type permutations for $r > r_{max}$ are of type B.
  In some cases, due to the particular topological properties of the network we
can establish the values r_0 and r_{max}. For example, for four-dimensional hypercube
network H_4 (see Sect. 2.2), it is easy to establish that for strong connectivity of nodes
in set T_1 we need at least six edges in *up*. Therefore, $r_0 = 5$.
  Below we present a sketch of an algorithm for estimating both the ternary and TIM-
spectra, which does not involve preliminary information about network's topology.
Introduce the following counters:

M_{ir}—the counter for the event "the anchor of r-permutation equals i";
$M_{ijr}(up)$—the counter for the event "the anchor of r-permutation equals i and the
component j is *up*";
$M_{ijr}(mid)$—the counter for the event "the anchor of r-permutation equals i and the
component j is *mid*";
$M_{ijr}(down)$—the counter for the event "the anchor of r-permutation equals i and
the component j is *down*".

Estimating the ternary D-spectrum and TIM spectra

(a) Put initially $M_{ir} = 0$, $M_{ijr}(up) = 0$, $M_{ijr}(mid) = 0$, $M_{ijr}(down) = 0$.

(b) For each $r = 0, 1, 2, \ldots, n - 1$ **DO** the following:

(b-1) Simulate a random permutation π of network components. Turn into up the first r components in π, and the remaining $(n - r)$—into state mid. (This is a simulated replica of an r-permutation).

(b-2) Find out the anchor $i = \delta(\pi_r)$ of the permutation. (This is carried out by a sequential destruction of components in mid and system state check after a component in mid is turned into $down$)

(b-3) Put $M_{ir} := M_{ir} + 1$; For each component j, $j = 1, \ldots, n$ do the following:

If j is in up state then put $M_{ijr}(up) := M_{ijr}(up) + 1$.
If j is in mid state then put $M_{ijr}(mid) := M_{ijr}(mid) + 1$.
If j is in $down$ state then put $M_{ijr}(down) := M_{ijr}(down) + 1$.

(c) Repeat steps (b-1)–(b-3) M times.

(d) For each $r = 1, \ldots, n - 1$, $i = 0, \ldots, n - r$, compute $\widehat{f}_r(i) = \frac{M_{ir}}{M}$. From this obtain the estimated values of the ternary D-spectrum:

$$\widehat{F}_r(x) = \sum_{j=0}^{x} \widehat{f}_r(j);$$

For each $r = 1, \ldots, n-1$, $i = 0, \ldots, n-r$, $j = 1, \ldots, n$, compute the following fractions:

$$\widehat{f}_r(i|j - up) = \frac{M_{ijr}(up)}{M}, \quad \widehat{f}_r(i|j - mid) = \frac{M_{ijr}(mid)}{M},$$

$$\widehat{f}_r(i|j - down) = \frac{M_{ijr}(down)}{M}.$$

From this, obtain the estimated values of TIM-spectra, according to Definitions 1.5.2-b. #

The most CPU time of the above procedure is consumed on checking system $DOWN$ state after each destruction step. Essential acceleration here can be achieved by appropriate use of so-called DSS—disjoint set structures, see [4], page 30. Another way to accelerate finding the anchor in an r-permutation, is to use the well known *bisection*, see [7].

The question of principal interest is the choice of M. It depends essentially on the number of components n and the desired estimation accuracy. For practical purposes estimation of system $DOWN$ probability with relative error 0.2–0.4 % is quite satisfactory. We see no other way than carrying out a series of experiments in which M is varied to achieve the desired level of accuracy.

It is worth noting that having $M = 100,000$ for estimating each $F_r(x)$ in the H_4 network, consumes only about 8 s of CPU time and guarantees the desired accuracy.

References

1. Barlow RE, Proschan F (1975) Statistical theory of reliability and life testing. Rinehart and Winston Inc., Holt
2. Birnbaum ZW (1969) On the importance of different components in multicomponent system. In: Krishnaiah PR (ed) Multivarite analysis-II, Academic Press, New York, pp 581–592
3. Elperin T, Gertsbakh IB, Lomonosov M (1991) Estimation of network reliability using graph evolution models. IEEE Trans Reliab 40(5):572–581
4. Gertsbakh I, Shpungin Y (2009) Models of network reliability: analysis, Combinatorics and Monte Carlo. CRC Press, Boca Raton
5. Gertsbakh I, Shpungin Y (2011) Network reliability and resilience. Springer Briefs in Electrical and Computer Engineering, Springer, New York
6. Gertsbakh I, Shpungin Y (2012) Combinatorial approach to computing importance indices of coherent systems. Probab Eng Inf Sci 26:117–128
7. Gertsbakh I, Rubinstein R, Shpungin Y (2014) Permutational methods for performance analysis of stochastic flow networks. Probab Eng Inf Sci 28:21–38
8. Gertsbakh I, Shpungin Y (2012) Failure development in a system of two connected networks. Transp Commun 13(4):255–260
9. Gertsbakh I, Shpungin Y (2012) Stochastic models of network survivability. Qual Technol Quant Manag 9(1):45–58
10. Samaniego FJ (1985) On closure under ifr formation of coherent systems. IEEE Trans Reliab 34:69–72
11. Samaniego FJ (2007) System signatures and their application in engineering reliability. Springer, New York

Chapter 2
Applications

Abstract Sections 2.1–2.4 present numerical illustrations and applications of the theory developed in Chap. 1. Section 2.1 presents reliability calculations for H_4 network. The network has 16 nodes, 32 edges and two sets of terminals, T_1 and T_2. An edge $e = (a, b)$ in state *up* provides high communication speed between a and b. If this edge is in state *mid*, the $a \leftrightarrow b$ communication goes with reduced speed; *down* state for an edge means that this edge is erased. Edge state is chosen randomly and independently according to probabilities p_2, p_1, p_0 for *up*, *mid* and *down* state, respectively. System *UP* state is defined as the existence of high speed communication between nodes of T_1 and existence of a path of operational edges between any pair of nodes of T_2. We present data on network reliability and characterize the ternary D-spectrum numerically and graphically. Section 2.2 considers a stochastic source-terminal flow network problem for a dodecahedron network. In this network, an edge $e = (e, b)$ is a *pair* of directed links for $a \rightarrow b$ and $b \rightarrow a$ flows. Each link has capacity 6, 3, or 0 for *up*, *mid* and *down* state, respectively. The network has two *DOWN* states, *DOWN*1 and *DOWN*2, for the flow less than L_1 or L_2, respectively $(L_1 > L_2)$. We present data on network reliability for various values of edge state probabilities $\mathbf{p} = (p_2, p_1, p_0)$. Section 2.3 is an example of a rectangular grid network with 100 nodes and 180 edges. Components subject to failure are the nodes. If a node is *down* all edges adjacent to it are erased and the node gets isolated. If a node is in *mid* state, it has only horizontal or vertical edges, depending on the position of the node. For this network we calculate the probability that the largest connected node set (an analogue to a "giant" component) has less than L nodes. Section 2.4 analyzes edges importance data for H_4 network. Section 2.5 deals with networks having independent and *nonidentical* components. Component i is in state *up, mid* and *down*, with probability $p_2^{(i)}, p_1^{(i)}$ and $p_0^{(i)}$, respectively. In this situation, different failure sets with the same number of components in *up, mid, down* have different probabilistic weights and it is not possible to use the ternary spectrum technique. For calculating network reliability, we present an efficient and accurate Monte Carlo method based on a modification of M.V. Lomonosov's evolution algorithm [3, 4]. Its action is illustrated by examples of a flow network and a grid network.

I. Gertsbakh et al., *Ternary Networks*, SpringerBriefs in Electrical
and Computer Engineering, DOI: 10.1007/978-3-319-06440-6_2,
© The Author(s) 2014

Keywords Dodecahedron network · Hypercube network · Flow-network · Grid network · Giant component · Component importance · Nonidentical components · Evolution algorithm

2.1 Hypercube Network

We consider fourth order hypercube H_4. It has 16 nodes and 32 edges, see Fig. 2.1. Each edge, independently of other edges, can be in three states: *up*, *mid* and *down* with respective probabilities p_2, p_1 and p_0, $p_2 + p_1 + p_0 = 1$. States *up* and *mid* provide high and medium connection speed, respectively. Edge *down* state means loss of connection. We say that node set T_1 is *strongly connected* if any pair of nodes from this set is connected by a path consisting only of edges providing high speed connection. We say that node set T_2 is *weakly connected* if any pair of nodes in this set is connected by a path of operational edges. We define *UP* state of our system as the presence of strong connection between nodes $T_1 = \{0, 3, 7, 12\}$ *and* weak connection between nodes of $T_2 = \{1, 4, 6, 15\}$, see Fig. 2.1. System *DOWN* state is absence of strong connection for T_1 *or* absence of weak connection for T_2, or both.

Our main tool for computing $P(DOWN)$ is the ternary D-spectrum. For this purpose we used Monte Carlo procedure, the algorithmic details of which were described in Sect. 1.6. Now let us mention some specific properties of the ternary spectrum in this example. First, it is easy to check from Fig. 2.1 that the minimal number of edges providing strong connection for T_1 equals 6. Therefore, if we have $r \leq 5$ edges *up* and other edges in *mid*, the system is *DOWN*. Thus, the ternary spectrum starts with $F_6(x), x = 1, 2, ..., 26$. In the course of simulation it is revealed that $r_{max} = 28$, i.e. if 29 or more edges are *up*, the system can not fail. Therefore the ternary spectrum is the collection $F_6(x), F_7(x), ..., F_{28}(x)$. Figure 2.2 presents a sample of graphs of $\{F_i(x)\}$.

Data on system *DOWN* probability is presented in Table 2.1. We see that for $p_2 \leq 0.5$ and p_1 in the range $[0.1, 0.4]$, the system has low reliability—$P(DOWN) \geq 0.4$. To provide reliability of about 0.85 or higher, p_2 should exceed 0.6 and p_1 must not exceed 0.15.

We will often characterize the accuracy of estimating *DOWN* probability by so-called relative error (RE). It is calculated as follows. The whole simulation experiment based on M replications for estimating the ternary spectrum and for calculating $P(DOWN) = \ell$ is repeated 10 times. Let the result of the i-th experiment be denoted by ℓ_i. Denote by S the estimate of the standard deviation computed as

$$S = [\sum_{i=1}^{10} (\ell_i - \ell_0)^2 / 9]^{0.5}, \tag{2.1}$$

where $\ell_0 = \sum_{i=1}^{10} \ell_i / 10$. Then the RE is defined as

Fig. 2.1 Hypercube of order four. $T_1 = \{0, 3, 7, 12\}$, $T_2 = \{1, 4, 6, 15\}$

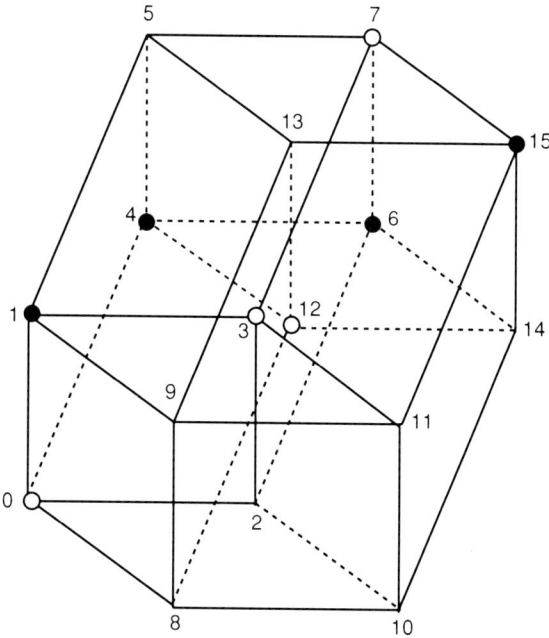

$$RE = \frac{S}{\ell_0}. \qquad (2.2)$$

Our calculations reveal that for $M = 100{,}000$ and $p_2 = 0.6, p_1 = 0.3$, the $RE \approx 0.01$. It means that the absolute error in estimating the *DOWN* probability is about $\pm$**0.002**. For $p_2 = p_1 = 0.3$, $P(DOWN)$ is close to 1, and the $RE \approx 0.0003$, which means that the absolute error is about $\pm$**0.0003**.

2.2 Flow Network

The network in this example is a dodecahedron graph shown on Fig. 2.2. Nodes 1 and 10 are the source and the sink, respectively. The edges incident to the source $s = 1$ allow the flow to be directed only outward as $(s \mapsto 2)$, $(s \mapsto 16)$, $(s \mapsto 5)$, and all edges incident to sink $t = 10$ allow the flow to go only in the direction of the sink. All other edges shown on Fig. 2.2, represent a *pair* of independent links, allowing the flow to go in opposite directions. For example, there are *two* directed edges connecting nodes 2 and 15: $(2 \mapsto 15)$ and $(15 \mapsto 2)$. There are in total $30 + 24 = 54$ edges in this flow network (Fig. 2.3).

The components subject to failure are the edges. All edges can be in three states: *up, mid,* and *down.* An edge in *up* has maximal flow capacity 6. An edge in *mid*

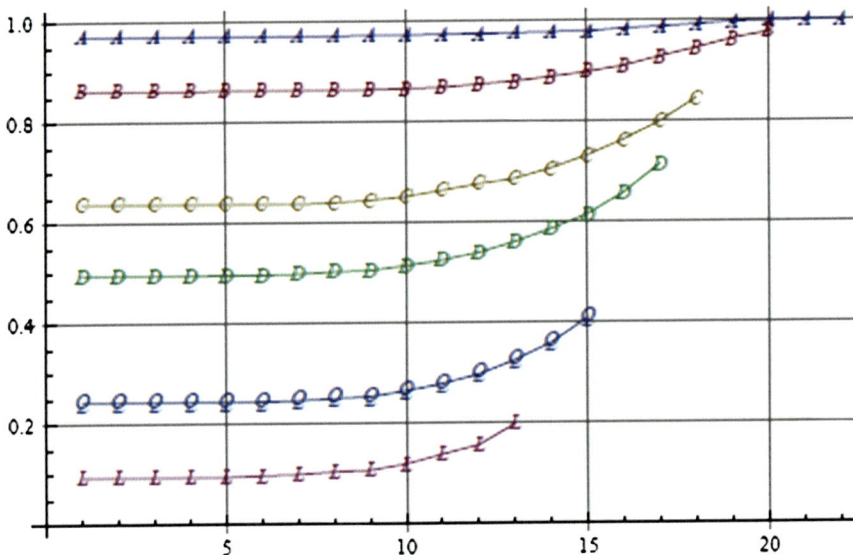

Fig. 2.2 Sample of $F_i(x)$. From *top* to *bottom*: $F_{10}, F_{12}, F_{14}, F_{15}, F_{17}, F_{19}$

Table 2.1 Network *DOWN* Probability $P(DOWN)$

p_1	$p_2 = 0.1$	$p_2 = 0.2$	$p_2 = 0.3$	$p_2 = 0.4$	$p_2 = 0.5$	$p_2 = 0.6$	$p_2 = 0.7$	$p_2 = 0.8$
0.1	0.9999	0.9976	0.9563	0.7790	0.4661	0.1887	0.0500	0.0073
0.2	0.9999	0.9957	0.9409	0.7412	0.4250	0.1660	0.0437	–
0.3	0.9959	0.9940	0.9312	0.7299	0.4099	0.1586	–	–
0.4	0.9928	0.9928	0.9267	0.7161	0.3981	–	–	–

state has maximal flow capacity 3, and an edge in *down* state has zero capacity (does not exist). The "parallel" edges, like (5, 6) and (6, 5) are *independent* with respect to their state. So, it may happen that edge (5, 6) will have capacity 6, and edge (6, 5)—capacity 3.

Obviously, the maximal flow equals 18. We will introduce two threshold values for the random flow $L_2 = 4.5$ and $L_1 = 10$ and define the following *DOWN* states for the network:

$$DOWN2 : \text{maximal flow } MF \leq 4.5,$$

$$DOWN1 : \text{maximal flow } MF \leq 10.$$

Suppose that all edges of the network are *up*, and therefore the network is in *UP* state and the $MF = 18$. Let edges start to fail in random order. There will be an instant when the MF will become smaller or equal 10. In that case we say that the network entered state *DOWN1*. Suppose the edges continue to fail and the MF becomes ≤4.5. Then, by our definition, the network will be in state *DOWN2*. So, the $MF \in [18, 10)$

Fig. 2.3 The dodecahedron
network. Nodes 1 and 10
are the source and the sink,
respectively. All edges except
incident to 1 and 10, represent
a pair of independent edges

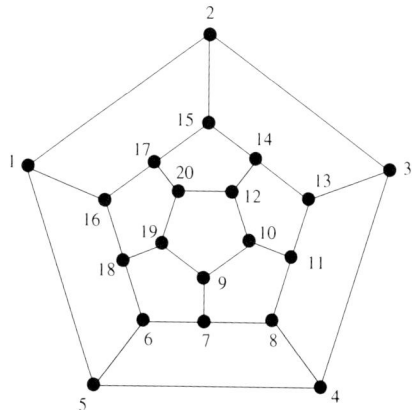

means the *UP* state, $4.5 < MF \leq 10$ means *DOWN1* state, and for $MF \leq 4.5$ the
network is in *DOWN2* state.

We just described a ternary system which is not binary: it can be in three states.
This extension does not complicate the formal part of our exposition. In this new
situation we have not a single ternary D-spectrum, but *two* ternary D-spectra: one for
the state *DOWN1* and another—for state *DOWN2*. When calculating one of these
spectra, we can completely ignore the presence of another.

Let us describe details of the simulation of the above spectra in details. For each
r we generated $N = 100{,}000$ permutations. It guarantees quite good accuracy in
computing network *DOWN* probabilities. For example, the estimate of $P(DOWN1)$
for $p_2 = 0.6, p_1 = 0.3, p_0 = 0.1$ is

$$P(DOWN1) = \mathbf{0.4407}$$

with relative error (RE) is about 0.002. It corresponds to an absolute error $\delta \approx$
0.0009. We calculated the same probability using crude Monte Carlo (CMC) based
on 1,000,000 replications and the result was

$$P_{CMC}(DOWN1) = \mathbf{0.4416}.$$

Table 2.2 presents the *DOWN1* probabilities for a sample of seven different combinations of p_2, p_1, p_0 probabilities.

We see from this table that even the "most reliable" combination $p_2 = 0.7, p_1 =$
0.2, $p_0 = 0.1$ has rather large value of *DOWN1* probability:

$$P(DOWN1) = \mathbf{0.3243}.$$

Table 2.2 Flow network
DOWN1 and *DOWN2*
probabilities

p_2	p_1	p_0	$P(DOWN1)$	$P(DOWN2)$
0.7	0.2	0.1	0.3243	0.0021
0.6	0.2	0.2	0.7042	0.1481
0.5	0.2	0.3	0.9274	0.4384
0.4	0.2	0.4	0.9908	0.7556
0.6	0.3	0.1	0.4407	0.2691
0.5	0.3	0.2	0.8064	0.1890
0.4	0.3	0.3	0.9654	0.5110

Fig. 2.4 $F_5(x)$ for $MF \leq 4.5$
(*lower curve*) and $MF \leq 10$
(*upper curve*)

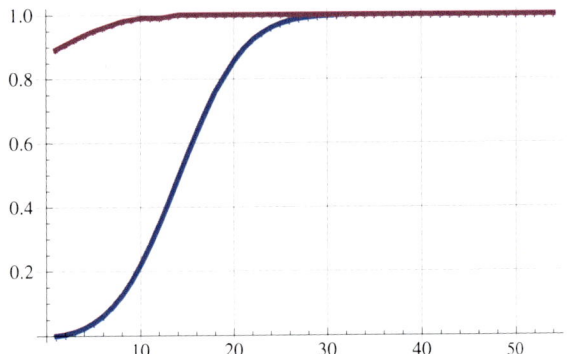

If we change the *DOWN* definition to *DOWN2* meaning that $P(DOWN2) = P(MF \leq 4.5)$, the network becomes considerably more reliable, as we see from the right column of the Table.

The Monte Carlo estimation of ternary D-spectra for $MF = 4.5$ and $MF = 10$ (with 100,000 replications) took about 180 s of CPU time. It should be noted that the search of the anchor position in a permutation was made by using bisection algorithm, see [8].

For $MF = 10$, $r = 0$ and $r = 1$, the network is always *DOWN*. (This was called in Sect. 1.2 as existence permutations of type A). We remind that for no components in *up* and all components in *mid*, or only one component in *up* and remaining in *mid*, the network is always *DOWN*. It is easy to understand it: if all edges are in *mid*, the maximal flow can not exceed 9; the same is true if only one of the edges is in *up*. It turns out that $r_{max} = 52$, i.e. when 53 edges are *up*, the system is always *UP*. The upper curves on Figs. 2.4 and 2.5 show the graphs of $F_5(x)$ and $F_{21}(x)$ for $MF = 10$.

For $MF = 4.5$ the principal behavior of the $F_r(x)$ curves is similar, see the lower graphs on Figs. 2.4 and 2.5. Contrary to the previous case, permutations of type A do not exist here. Permutations of type B appear for $r = 52$ and $r = 53$.

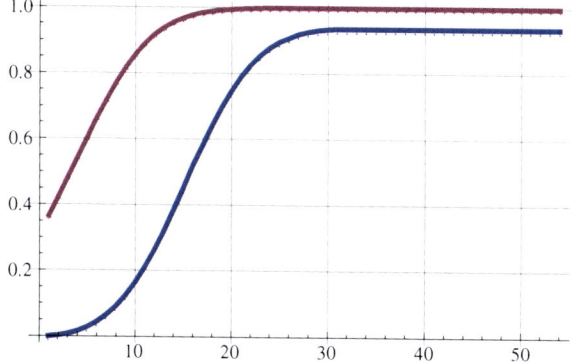

Fig. 2.5 $F_{21}(x)$ for $MF \leq 4.5$ (*lower curve*) and $MF \leq 10$ (*upper curve*)

Fig. 2.6 Rectangular grid with 100 nodes and 180 edges. Bold nodes are *down*. Nodes marked by x are in *mid*

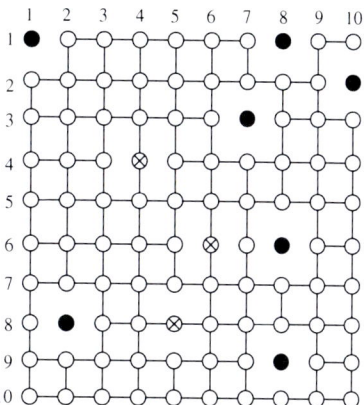

2.3 Survival of a Ternary Grid Network

In this section we will consider a 10×10 grid network with nodes having three states:

(i) *down*—all four edges adjacent to this node are erased.
(ii) *mid*—erased are two out of four edges adjacent to the node. Node is called *even* if the sum of its row and column numbers in the grid are even. Otherwise it is called *odd*. At even node in *mid*, the adjacent *horizontal* edges are erased; at odd node—vertical adjacent edges are erased.
(iii) *up*—all edges incident to this node are initially intact. These edges, however may be erased if neighboring nodes are in *down* or in *mid*.

Figure 2.6 shows the rectangular grid with some nodes in *down* and *mid*.

Table 2.3 Fragments of ternary spectrum $F_r(x)$ for $L_1 = 80$

r	$x = 0$	$x = 5$	$x = 10$	$x = 15$	$x = 20$
0	1	1	1	1	1
10	1	1	1	1	1
20	0.9999	0.9999	1	1	1
30	0.9755	0.9910	0.9994	1	1
40	0.7171	0.8218	0.9553	0.9987	1
45	0.4913	0.6403	0.8400	0.9842	1
50	0.2805	0.3798	0.6326	0.9465	1
60	0.0514	0.0839	0.1954	0.6353	1
70	0.0035	0.0074	0.0274	0.2121	1
80	0.0000	0.0001	0.0021	0.0358	1
90	0	0	0	0	0

Definition 2.3.1 Network component.

A subset of nodes V_0 is called a *component* of the network if there is at least one path from each node $v \in V_0$ to each other node in V_0 and no other node in $V \setminus V_0$ can be added to V_0 while preserving this property. #

Network proper functioning, e.g. information delivery, power supply etc can be guaranteed only if its *maximal size* component is large enough [7, 11]. Accordingly, we will define network *UP* state if its maximal component exceeds some critical level $L_{\max}$. We will consider two values for this level: $L_{\max} = L_1 = 80$ and $L_{\max} = L_2 = 50$. Accordingly, we define two versions of network *UP* state:

$$UP1 : \text{maximal component size exceeds } L_{\max} = L_1 = 80$$

and

$$UP2 : \text{maximal component size exceeds } L_{\max} = L_2 = 50.$$

Correspondingly we have network *DOWN*1 state as the complement to *UP*1 and *DOWN*2 as the complement to *UP*2.

Let us consider the ternary spectra for the network. Table 2.3 shows a fragment of the spectrum for $L_1 = 80$, for a selection of r values (column 1) and x values. The ternary spectrum was calculated using Monte Carlo simulation with 100,000 replications for each r value. The total simulation time took 179.5 s of CPU time.

Let us remind that $F_r(x)$ presented in the Table 2.3 is the probability that the network is in *DOWN1* if r of its randomly chosen nodes are *up* and x of its randomly chosen nodes are in *down*, and therefore $(n - r - x)$ components are in *mid*. We see from the table that for $r = 0$ and $r = 10$ we have the situation described in Sect. 1.2 as permutations of type A: when r components are *up* and all other $n - r$ components are in *mid*, the network is in *DOWN1*.

For $r \geq 90$, network is in *UP1* for all x values. We see that for $x = 20$ the network is in *DOWN1* with probability 1, as it should be by the definition of *DOWN1*. It is interesting that for $r = 70$, $P(DOWN1)$ changes from 0.0274 to 1 in a narrow

Table 2.4 *DOWN1* probability for grid network

p_2	p_1	p_0	$P(DOWN1)$
0.7	0.2	0.1	**0.0920**
0.4	0.2	0.4	1
0.6	0.3	0.1	**0.2889**
0.5	0.3	0.2	0.9557
0.4	0.3	0.3	0.9999

Table 2.5 Fragments of ternary spectrum $F_r(x)$ for $L_2 = 50$

r	$x = 0$	$x = 5$	$x = 10$	$x = 15$	$x = 20$	$x = 30$	$x = 40$	$x = 50$
0	1	1	1	1	1	1	1	1
10	0.9999	0.9999	1	1	1	1	1	1
20	0.9819	0.9874	0.9956	0.9990	0.9998	1	1	1
30	0.7356	0.8104	0.9049	0.9622	0.9878	0.9996	1	1
40	0.2756	0.3379	0.5020	0.6752	0.8161	0.9719	0.9996	1
45	0.1323	0.1895	0.3279	0.4944	0.8090	0.9792	0.9999	1
50	0.0576	0.0084	0.1679	0.2906	0.4418	0.7848	0.9835	1
60	0.0081	0.0120	0.0293	0.0625	0.1186	0.3688	0.8104	–
70	0.0005	0.0009	0.0026	0.0063	0.0141	0.8734	–	–
80	0.0000	0.0000	0.0004	0.0006	0.0006	–	–	–
85–99	0	0	0	0	0			

interval $x \in [10 - 20]$, which reminds the phase transition phenomena in percolation. Table 2.4 presents *DOWN1* probabilities for five different combinations of p_2, p_1, p_0.

The results shown in bold were checked by crude Monte Carlo (CMC) simulation with 1,000,000 replications. For example, the CMC estimate is 0.09198, i.e. differs from our result by 0.00004.

Table 2.5 presents the ternary spectrum for $L_{max} = L_1 = 50$. Similarly to the previous case, for $r = 0, ..., 9$, all $F_r(x)$ values are equal 1. The *DOWN2* probabilities are smaller than the corresponding values $P(DOWN1)$, as expected.

For r in the interval 85–99, we have $F_r(x) = 0$ (permutation of type B) which means that the network is always *UP2*. $F_{70}(x)$ has a slow increase of $0.0005 - 0.0141$ when x changes from 0 to 20 and a very rapid increase to 0.8734 when x approaches 30. It also reminds a phase transition. Table 2.6 presents a collection of $P(DOWN2)$ probabilities for a selection of p_2, p_1, p_0 values.

It is seen that to provide $P(DOWN2) < 0.1$, the probabilities for a node to be in *up* and *mid* respectively, should be close to $(0.7, 0.2)$. For that case the average number of *up* nodes will be near 70, and *down* nodes—around 10. This is in agreement with data shown for $r = 70$ in Table 2.5.

Table 2.6 Grid network
DOWN2 probability
P(DOWN2)

p_2	p_1	p_0	P(DOWN2)
0.7	0.2	0.1	0.0073
0.6	0.2	0.2	0.1738
0.5	0.2	0.3	0.7236
0.4	0.2	0.4	0.9862
0.6	0.3	0.1	0.0520
0.5	0.3	0.2	0.4598
0.4	0.3	0.3	0.9297

2.4 Component Importance in Hypercube H_4 Network

Let us remind that network UP state was defined as the presence of strong connectivity
of $T_1 = \{0, 3, 7, 12\}$ and weak connectivity of $T_2 = \{1, 4, 6, 15\}$, see Fig. 2.1.
Table 2.7 presents the up-down edge importance indices computed by formula (1.5.3).
We remind also, that this formula represents the partial derivative of the reliability
function with respect to p_2 when edge mid probability p_1 remains unchanged (p_2
increases on account of p_0). This definition copies in fact the Birnbaum Importance
Measure (BIM) used in binary systems, see [1, 2, 6].

Three variants of **p** were considered: **p** = (0.6, 0.3, 0.1), **p** = (0.5, 0.3, 0.2) and
p = (0.4, 0.3, 0.3), see columns 3, 4, 5 in the table.

Figure 2.7 presents the data of this table in a form convenient for visual analysis
of component TIM's.

It is clearly seen that the edges have practically the same ranking for all three
versions of **p**. The most important edge is 11 which connects nodes of sets T_1 and
T_2. The second group of edges consists of 14 edges which have one of its ends in the
set T_1. These edges are shown by bold in Table 2.7. The importance values for this
category are in the range [0.074–0.100]. The third group constitute the remaining
edges with importance measures (for **p** = {0.6, 0.3, 0.1}) in the range [0.013–0.034].

If we think about replacing an edge by a more reliable one, the best candidate
are edge 11 and edges in the second group. If, for example, we replace edge 11
and 15 having $p_2 = 0.6$ by an edge with $p_2 = 0.7$ (and $p_0 = 0$, respectively), we
might expect reliability increase by $0.1 \cdot (0.150 + 0.097) \approx 0.025$, or equivalently,
P(DOWN) decrease by the same quantity. As it is seen from Table 2.1, for $p_2 =
0.6, p_0 = 0.3$, P(DOWN) = 0.1586. Therefore the above edge replacement will
bring P(DOWN) to ≈ 0.1336, which is a quite significant improvement.

Table 2.8 presents the down-mid component importance values (see Definition
1.5.1-B).

First of all, it is seen that the values in this table are significantly smaller than
the corresponding values in the previous table. This is expected, since the up-down
importance reflects the effect from reinforcing a component in up on the account
of reducing p_0. It is bigger than the mid-down importance calculated for similar
reinforcement of mid on the account of reducing p_0. Second conclusion is that the

Table 2.7 Edge *up-down* TIM's for H_4 by (1.5.3), $T_1 = (0, 3, 7, 12)$, $T_2 = (1, 4, 6, 15)$

i	Edge	$\mathbf{p} = (0.6, 0.3, 0.1)$	$\mathbf{p} = (0.5, 0.3, 0.2)$	$\mathbf{p} = (0.4, 0.3, 0.3)$
1	(0, 1)	0.084	0.145	0.143
2	(0, 2)	0.076	0.133	0.130
3	(0, 4)	0.091	0.152	0.142
4	(0, 8)	0.083	0.131	0.117
5	(1, 3)	0.092	0.144	0.122
6	(1, 5)	0.025	0.066	0.078
7	(1, 9)	0.032	0.067	0.067
8	(2, 3)	0.089	0.144	0.131
9	(2, 6)	0.028	0.061	0.055
10	(2, 10)	0.018	0.048	0.056
11	(3, 7)	0.150	0.252	0.237
12	(3, 11)	0.074	0.104	0.077
13	(4, 5)	0.024	0.062	0.077
14	(4, 6)	0.033	0.075	0.079
15	(4, 12)	0.097	0.149	0.133
16	(5, 7)	0.075	0.108	0.083
17	(5, 13)	0.025	0.053	0.052
18	(6, 7)	0.088	0.132	0.112
19	(6, 14)	0.035	0.071	0.072
20	(7, 15)	0.082	0.118	0.087
21	(8, 9)	0.029	0.061	0.070
22	(8, 10)	0.034	0.063	0.063
23	(8, 12)	0.091	0.140	0.121
24	(9, 11)	0.019	0.042	0.040
25	(9, 13)	0.025	0.050	0.050
26	(10, 11)	0.013	0.029	0.037
27	(10, 14)	0.022	0.048	0.046
28	(11, 15)	0.024	0.051	0.050
29	(12, 13)	0.082	0.116	0.090
30	(12, 14)	0.076	0.112	0.089
31	(13, 15)	0.027	0.048	0.039
32	(14, 15)	0.030	0.072	0.090

edge ranking remains practically the same for all three sets of *up, mid* and *down* probabilities, see Fig. 2.8.

If we check the edge positions in H_4, we will see that the largest values of *TIMs* have edges adjacent to the nodes of T_2. These edges are most important for providing connectivity of the nodes in this set (Fig. 2.8).

Remark Suppose we are interested in estimating the increase in system reliability achieved by replacing edge 5 having $p_2 = 0.5, p_1 = 0.3, p_0 = 0.2$ by an edge which has $p_2 = 0.8, p_1 = 0, p_0 = 0.2$. This is a replacement of type *mid-up* that leaves p_0

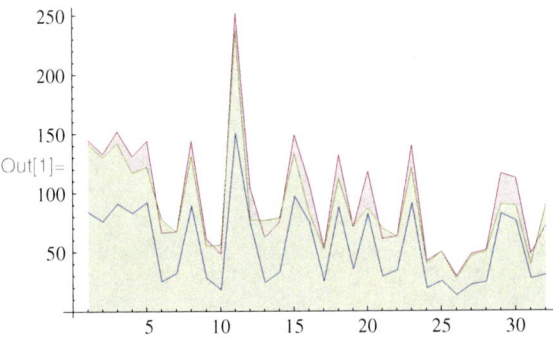

Fig. 2.7 *Up–down* TIM's for edges of H_4 multiplied by a factor 1000. Edge numbers are shown on the x-axis

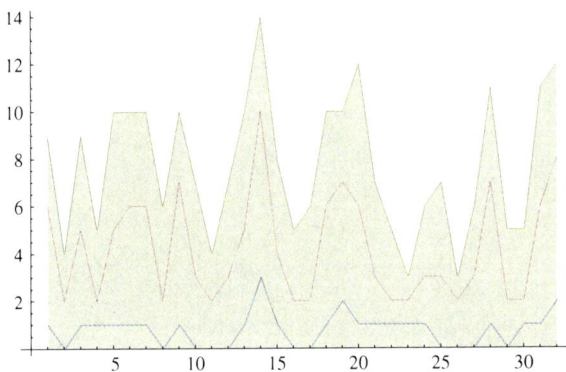

Fig. 2.8 *Mid–down* TIM's for edges of H_4 multiplied by a factor 1000. Edge numbers are on x-axis

unchanged. Following Definition 1.5.1-C, we see that we have to take the difference between *up-down* importance in Table 2.7 and *mid-down* importance in Table 2.8. We obtain for edge 5 and $p_0 = 0.2$ (column 4) the value $0.144 - 0.005 = 0.139$. For the increase of p_2 by $0.8 - 0.5 = 0.3$, we can expect the drop of *DOWN* probability by $\delta \approx 0.139 \cdot 0.3 = 0.0417$.

In general, estimation of the effect of component reinforcement on network reliability for ternary network is quite similar to the procedure described in [5], Chapter 2 for binary networks.

Table 2.8 Edge *down–mid* TIM's for H_4 by (1.5.4), $T_1 = (0, 3, 7, 12)$, $T_2 = (1, 4, 6, 15)$

i	Edge	$\mathbf{p} = (0.6, 0.3, 0.1)$	$\mathbf{p} = (0.5, 0.3, 0.2)$	$\mathbf{p} = (0.4, 0.3, 0.3)$
1	(0, 1)	0.001	0.006	0.009
2	(0, 2)	0.000	0.002	0.004
3	(0, 4)	0.001	0.005	0.009
4	(0, 8)	0.001	0.002	0.005
5	(1, 3)	0.001	0.005	0.010
6	(1, 5)	0.001	0.006	0.010
7	(1, 9)	0.001	0.006	0.010
8	(2, 3)	0.000	0.002	0.006
9	(2, 6)	0.001	0.007	0.010
10	(2, 10)	0.000	0.003	0.007
11	(3, 7)	0.000	0.002	0.004
12	(3, 11)	0.000	0.003	0.007
13	(4, 5)	0.001	0.005	0.010
14	(4, 6)	0.003	0.010	0.014
15	(4, 12)	0.001	0.004	0.008
16	(5, 7)	0.000	0.002	0.005
17	(5, 13)	0.000	0.002	0.006
18	(6, 7)	0.001	0.006	0.010
19	(6, 14)	0.002	0.007	0.010
20	(7, 15)	0.001	0.006	0.012
21	(8, 9)	0.001	0.003	0.007
22	(8, 10))	0.001	0.002	0.005
23	(8, 12)	0.001	0.002	0.003
24	(9, 11)	0.001	0.003	0.006
25	(9, 13)	0.000	0.003	0.007
26	(10, 11)	0.000	0.002	0.003
27	(10, 14)	0.000	0.003	0.006
28	(11, 15)	0.001	0.007	0.011
29	(12, 13)	0.000	0.002	0.005
30	(12, 14)	0.001	0.002	0.005
31	(13, 15)	0.001	0.006	0.011
32	(14, 15)	0.002	0.008	0.012

2.5 Evolution Process for Ternary Network

2.5.1 Lomonosov's Evolution Process with Merging

Let us remind that all calculations of network reliability (where the *UP* state was defined as the existence of strong and weak connectivity, or the existence of an $s - t$ flow, or as presence of largest component above critical size) were made under two important conditions: all components subject to failure are *identical* and stochastically *independent*. Below we will describe an ingenious Monte Carlo algorithm of

M.V. Lomonosov for estimating the probability of network connectivity which will allow to treat networks with independent components which have *different* failure probabilities. The first publication of this algorithm is dated 1991 [3], see also its detailed description in [4], Chap. 9.

Originally, Lomonosov's algorithms (LA) was designed for binary networks. We will first describe this version of LA for networks whose components subject to failures are binary edges. Later, in Sect. 2.5.2 we will describe an extension of the LA to ternary components.

Suppose that we have a network $\mathcal{N} = (V, E, T)$, where V is the node set, E is the edge set, and T is a subset of V of "special" nodes called *terminals*. The edges are binary: edge e_i is *down* with probability $q_i = 1 - p_i$ and *up* with probability p_i. Network *UP* state is defined as presence of terminal connectivity of all nodes $v \in T$.

The first idea of LA is to associate with each edge e_i an exponentially distributed random *birth time* τ_i with parameter λ_i having the following property. For an arbitrary chosen time value t_0

$$P(\tau_i \le t_0) = 1 - e^{-\lambda_i t_0} = p_i, \quad i = 1, 2, ..., n. \tag{2.3}$$

Let us assume that edge e_i is born at time τ_i and initially was in state *down*. At the instant τ_i of its birth it becomes *up* and stays in this state "forever". Then the probability that it will be "alive" at time t_0 equals $P(\tau_i \le t_0) = p_i$. For simplicity we put $t_0 = 1$ and note that from (2.3) follows

$$\lambda_i = -\ln q_i. \tag{2.4}$$

Now comes the following crucial observation: if we take a "snapshot" of the state of all edges at time instant $t_0 = 1$, we will see the network in the state which is stochastically equivalent to the static picture in which edge e_i is *up* or *down* with probability p_i or q_i, respectively. In particular, the snapshot will reveal the network in *UP* with probability identical to that obtained for the static edge "lottery".

The second idea of LA is to consider the edge birth times as a process developing in time. Let us consider an instructive example.

Example 2.5.1 Network with 5 nodes and 7 edges.
Figure 2.9 shows a network with 5 nodes and 7 edges. Network is *UP* if all its nodes are connected to each other. The birth process starts at $t = 0$ from state denoted as σ_0: no edges are born. Suppose that edge e_5 is born first. Note that the time ξ_1 of its birth $\xi_1 = \min\{\tau_1, \tau_2, ..., \tau_7\}$ and it is exponentially distributed with parameter $\Lambda_1 = \sum_{i=1}^{7} \lambda_i$:

$$\xi_1 \sim Exp(\Lambda_1).$$

The probability that the first birth will be of edge 5 equals λ_5/Λ_1.

The next edge born will be one of the remaining edges. Suppose that the next birth takes place at time instant $\xi_1 + \xi_2$. Due to the properties of exponential independent random variables, see e.g. [12], ξ_1 and ξ_2 are independent, and

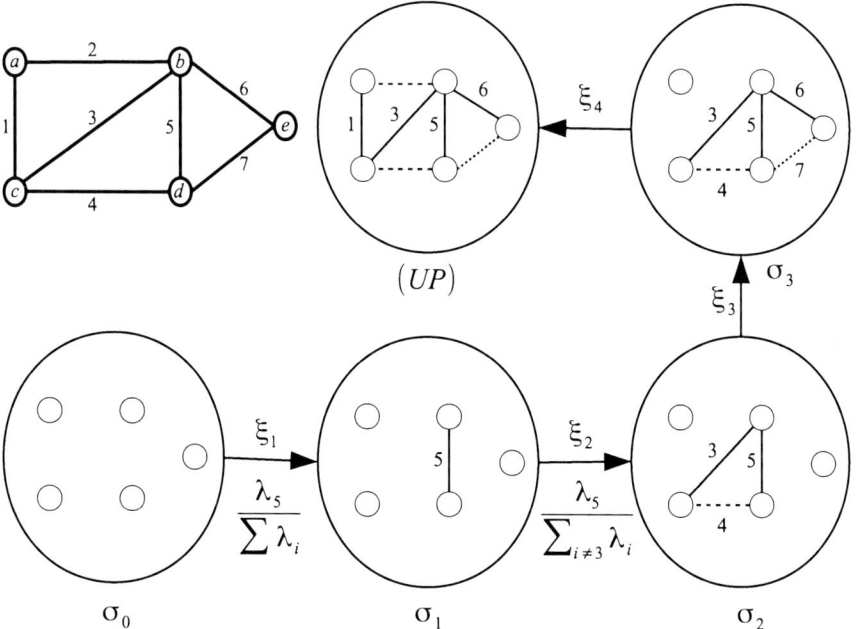

Fig. 2.9 Evolution trajectory

$$\xi_2 \sim Exp(\Lambda_2 = \sum_{i \neq 5} \lambda_i).$$

The probability that the second birth will be of edge 3 equals λ_3/Λ_2. The network enters now the state shown on Fig. 2.9 as σ_2. Continuing this process we will observe the third birth at the instant $\xi_1 + \xi_2 + \xi_3$ where

$$\xi_3 \sim Exp(\Lambda_3 = \sum_{i \neq 3,5} \lambda_i),$$

and ξ_3 is independent of ξ_1 and ξ_2. The third birth will be of edge 6 with probability λ_6/Λ_3. After the births of edges 5, 3 and 6, the network will be in state σ_3—see Fig. 2.9. Suppose that the next birth will be of edge 1. Suppose it happens at time instant $\xi_1 + \xi_2 + \xi_3 + \xi_4$. Obviously, all ξ_i-s are independent,

$$\xi_4 \sim Exp(\Lambda_4 = \sum_{i \neq 3,5,6} \lambda_i),$$

and fourth birth will be of edge 1 with probability λ_1/Λ_4. The birth of edge 1 signifies network entrance into the *UP* state. It is an absorbing state, and for all further births the system remains in *UP*.

The sequence of states $\omega = \{\sigma_0 \Rightarrow \sigma_1 \Rightarrow \sigma_2 \Rightarrow \sigma_3 \Rightarrow UP\}$ leading from the initial state σ_0 to the *UP* state is called *trajectory*. The probability that the trajectory will be as in our example is

$$\frac{\lambda_5}{\Lambda_1} \cdot \frac{\lambda_3}{\Lambda_2} \cdot \frac{\lambda_6}{\Lambda_3} \cdot \frac{\lambda_1}{\Lambda_4}.\#$$

Now let consider how the evolution process is used for computing network reliability. Denote the set of all trajectories ω by Ω. By conditioning over particular trajectory ω, we can represent the network *UP* probability in the following form:

$$P(\text{Network is } UP) = \sum_{\omega \in \Omega} P(\omega)P(\xi_1 + \xi_2 + \cdots + \xi_r \leq t_0 = 1|\omega). \qquad (2.5)$$

The probability $P(\xi_1 + \xi_2 + \cdots + \xi_r \leq t_0|\omega)$ can be computed in a closed form using hypo-exponential distribution, see [10, 12] and [4], Appendix B.

Let $\Lambda_1 > \Lambda_2 > \Lambda_3 > \cdots > \Lambda_r$, and $\xi_i \sim Exp(\Lambda_i)$ be independent. Then

$$P(\sum_{i=1}^{r} \xi_i \leq t) = 1 - \sum_{i=1}^{r} e^{-\Lambda_i t} \prod_{j \neq i} \frac{\Lambda_j}{\Lambda_j - \Lambda_i}. \qquad (2.6)$$

Since the quantity of interest in (2.5) is expressed as mathematical expectation, it can be estimated without bias as the sample average of computed probabilities $P(\xi_1 + \xi_2 + \cdots + \xi_r \leq t_0 = 1|\omega_j)$ over a sample of trajectories $\{\omega_1, \omega_2, ..., \omega_M\}$:

$$\widehat{P}(UP) = \frac{\sum_{j=1}^{M} P(\xi_{i_1} + \xi_{i_2} + \cdots + \xi_{r_j} \leq t_0 = 1)|\omega_j)}{M}. \qquad (2.7)$$

In the case when the components of the network are edges, the above described evolution method can be complemented by so-called *closure* or *merging* operation. Let us explain what means closure by Example 2.5.1. Consider the state σ_2 on Fig. 2.9. Edges 3 and 5 are already born. These edges connect three nodes: b, c and d. On each stage of the birth process the edges born and their nodes create components, i.e. connected sets of nodes. So, nodes c, b and d belong to one component. Note that edge 4 which connects nodes c and d of this component can be merged with the already born edges without changing the component which already has been born! More formally, we can conclude that edge $4 = (c, d)$ is not *relevant* and can be excluded from the further evolution process. After adding this edge, the remaining edges are only 1, 2, 6 and 7. Suppose that the next edge born will be 6, see state σ_3. Now the component has nodes b, c, d, e, and edge $7 = (d, e)$ joins the nodes belonging to the already existing component. So, now edge 7 can be added to the already born edges. After "merging" edge 7, we are left with only two non born edges, 1 and 2. The birth of any of them signifies the entrance into the *UP* state.

So, after each birth of an edge, we look for those edges, whose nodes belong to the already existing component. These edges are joined to this component and excluded from further considerations as irrelevant. Thus, after carrying out the closure operation, the trajectory leading from σ_0 to *UP* state represents a "thick" set, a "bundle" of trajectories. As the result, the dimension of the set of all trajectories Ω becomes smaller, and the estimate (2.7)—less variable.

Let us note that the estimate $P(DOWN) = 1 - P(UP)$ by (2.7) is free of so-called rare event phenomenon (unlimited increase of variance when $P(DOWN) \rightarrow 0$). In addition, the relative error of the estimator of $P(DOWN) = 1 - P(UP)$ is uniformly bounded with respect to the λ_i values. For additional details see [4], Chap. 9.

2.5.2 Extension of LA to Multistate Components

Originally, Lomonosov's algorithm described in [3] dealt with systems whose components subject to failure had only two states: *up* and *down*, or *exist-erased*. Now suppose that we have a network whose edges can be in three states: *up*, *down* and *mid*. In practice, edge $e = (a, b)$ in *mid* may correspond to a connection between a and b with reduced capacity or reduced speed of information exchange.

Suppose that we have a network whose components subject to failures are edges $e_1, e_2, ..., e_n$, and these edges are independent but *not* identical. Edge e_i has probabilities $p_2(e_i)$, $p_1(e_i)$ and $p_0(e_i)$ to be in state *up*, *mid* or *down*, respectively. Our goal is to find network static *DOWN* probability.

Consider a two-stage edge birth process: on the first stage the edge is born without specifying its state, *mid* or *up*. Afterwards, on the second stage, the born edge is assigned its "type"—*mid* or *up*. Let the probability that edge is born be

$$P(e \text{ born}) = p_b(e) = p_2(e) + p_1(e). \tag{2.8}$$

On the second stage, after the edge e is born, we carry out a *random lottery*, in which we assign state *up* to e with probability

$$P(\text{edge } e \text{ is } up | e \text{ is born}) = \frac{p_2(e)}{p_1(e) + p_2(e)} \tag{2.9}$$

and state *mid* with probability

$$P(\text{edge } e \text{ is } mid | e \text{ is born}) = \frac{p_1(e)}{p_1(e) + p_2(e)}. \tag{2.10}$$

Now let us associate with edge e its *birth time* $\tau_e \sim Exp(\lambda_e)$, where λ_e is chosen in such a way that the probability that at t_0 edge e is born equals $p_b(e)$:

$$P(\tau_e \le t_0) = 1 - e^{-\lambda_e t_0} = p_b(e). \tag{2.11}$$

A "snapshot" at t_0 on all edges (born or not) will reveal a picture which is sto-chastically equivalent to static independent "lotteries" which determine the edge states according to the above probabilities $\mathbf{p}(e) = (p_2(e), p_1(e), p_0(e))$. Now we can implement the evolution process of LA and simulate, similarly to the above described procedure, the ω trajectories.

Here, however, arises a complication. The principal property of the evolution process for a binary system was the fact that each trajectory starting from the initial state with no edges born led eventually to system UP state. This was guaranteed by the property of monotone systems: if all components are born, then the system should be UP. But now, we are in a situation where component can be born as mid or as "up". And we have no guarantee that even a full trajectory with all edges born brings the system into UP. Suppose, for example, that network UP state means the following: (i) all nodes are connected with each other by mid or up edges; (ii) nodes a, b, c must be "strongly" connected, i.e. connected only by edges in up. Suppose it happens that a full trajectory has only mid edges, i.e all births produced mid state. Then the trajectory can not end in the UP state, and it ends, in fact, in $DOWN$ state.

We will adopt the following zero-rule for the evolution process for which trajectories may end in $DOWN$ state:

If the trajectory $\omega^* = \{\sigma_0 \Rightarrow \sigma_1 \Rightarrow \cdots \sigma_r \equiv DOWN\}$ ends in $DOWN$, put

$$P(\xi_1 + \cdots + \xi_r \le t_0 | \omega^*) := 0. \tag{2.12}$$

This rule can be justified by the following reasoning. Supply our original network with a set of "superedges" (or "supercomponents"), such that two properties will be guaranteed:

(i) these edges have no mid state, and if they are born they always are in up;
(ii) If all superedges are born, network is always UP.

For example, network is UP if nodes s and t are connected only by edges in up and all other nodes are connected to each other. Then it will be enough to have one superedge $e_0 = (s, t)$. Another example is a flow network which is in UP if $s - t$ flow exceeds γ. Then add to the network $s - t$ superedge and assign to it capacity that exceeds the maximal flow L.

Now modify the original problem by adding the superedges (supercomponents) to the network and giving them negligible small up probability ε.

Let us see what will be the consequences of that. Suppose $\varepsilon = 10^{-10} = p_2(e_0)$ is the probability that superedge e_0 is born. (Remind that the superedge after its birth is in state up.) Then the probability that it will be born before $t_0 = 1$ equals

$$P(\tau_{e_0} \le 1) = 10^{-10} = 1 - e^{-\lambda}.$$

Therefore $\lambda \simeq 10^{-10} = \varepsilon$. This means that τ_{e_0} has a huge mean value $1/\varepsilon$. Therefore the probabilities of type $P(\xi_1 + \cdots + \xi_r + \cdots + \tau_{e_0} \le 1 | \omega)$ will be very close to zero. On the other hand, the presence of very small ε in the expressions of type $\lambda_j/(\lambda_1 + \cdots + \lambda_k + \varepsilon)$ will have no influence on the evolution process and the choice of the next born edge. Therefore, with probabilities arbitrary close to 1, the modified birth process will be developing exactly as if the superedges do not exist. After all "regular" edges will be born and only superedges are left, the evolution trajectories will continue by including the superedges. This will guarantee that all trajectories end in UP state. But the contribution to the numerator of (2.7) coming from these trajectories will be negligible. All this justifies the above "zero-rule" (2.12).

Now we are ready to formulate in general terms the Monte Carlo algorithm for estimating reliability of a ternary network with statistically independent and *non-identical* components.

Modified LA

(I) Using the $p_2(e), p_1(e), p_0(e)$, assign to each component e its birth rate λ_e and the type *mid* or *up* in case component e is born, according to (2.8)–(2.10).

(II) Generate the evolution trajectory by choosing on each step a component e from the set of nonborn components with probability λ_e/Λ_e, where Λ_e is the sum of λ's for all nonborn components. If component e is born, remember its birth time $\tau_e \sim Exp(\lambda_e)$. If closure operation is used, apply it, and merge the closing components to the existing ones and cross them out from the list of non born components.

After each birth, check if the trajectory has reached the UP state. If the UP state is being reached, STOP, and go to (III); otherwise continue the birth process until all non born components are exhausted. If UP state is not being reached remember the corresponding trajectory and mark it as "zero" type.

(III) For each trajectory ω of non "zero" type compute $P(\xi_1 + \xi_2 + \cdots + \xi_r \le 1|\omega)$ by (2.5)–(2.6); put $P(\xi_1 + \xi_2 + \cdots + \xi_r \le 1|\omega) = 0$ for zero type trajectory.

(IV) Put $P(DOWN) := P(DOWN) + P(\xi_1 + \xi_2 + \cdots + \xi_r \le 1|\omega)$

(V) Repeat Steps (II–IV) M times. Put $P(DOWN) := P(DOWN)/M$.

2.5.3 Flow Network

Let us consider the dodecahedron flow network studied earlier in Sect. 2.2. In Sect. 2.2 it was assumed that all edges have identical probabilities to be in *up, mid* or *down*. Now we introduce different probabilities for the edges to fail, according to the following rule. Nodes are numbered $j = 1, 2, ..., 20$. Edge $e = (i, j)$ is declared even if $i + j$ is even, and odd—otherwise. Assume that even edges have probabilities $p_2(a), p_1(a), p_0(a)$, and odd—$p_2(b), p_1(b), p_0(b)$. Edge capacities remain the same: 0, 3, and 6 for *down, mid* and *up* edge, respectively. Similar to the example in Sect. 2.2. there are two threshold values for the $s - t$ flow $L_1 = 10$ and $L_2 = 4.5$.

To find network $DOWN$ probability, we applied the modified LA. The closure operation is not applicable here since the UP criterion is not formulated in terms of node connectivity. The number of simulated trajectories was $M = 1,000,000$. All

Table 2.9 $P(DOWN)$ for flow network with two types of edges

i	$p_2(a); p_2(b)$	$p_1(a); p_1(b)$	$p_0(a); p_0(b)$	L	$P(DOWN)$	RE	CPU sec
1	0.6; 0.4	0.3; 0.4	0.1; 0.2	10	**0.757**	0.1 %	123
2	0.6; 0.4	0.3; 0.4	0.1; 0.2	4.5	**0.119**	0.2 %	103

simulation results obtained by the evolution method were checked by CMC with 10^6 replications, and numerically the results coincide with the results obtained by the LA with high accuracy (Table 2.9).

Below is a short summary of simulation results using LA:

Line $i = 1$ presents the *DOWN* probability for $L_1 = 10$. The first figure for p_2, p_1, p_2 is for even edges, the second— for odd edges. Line 2 shows similar results for reduced maximal flow $L_2 = 4.5$. Network *DOWN* changes considerably with the reduction of the maximal flow. The RE's are rather small and the CPU time is quite satisfactory. Let us note that the zero-type trajectories were observed in about 8 % of replications for $L_1 = 10$ and were not observed for $L_1 = 4.5$.

2.5.4 Grid Network

Here we consider a 10×10 grid network shown on Fig. 2.5. The network has 100 nodes. Node failure (*down* state) means erasing all edges adjacent to this node. If node v is in *mid* state, and this node is even (row number + column number is even), then the horizontal edges adjacent to v are erased. If node v is in *mid* and its row number + column number is odd, then the vertical edges adjacent to v are erased.

Node state probabilities are

$$p_{up}(a) = 0.6, p_{mid}(a) = 0.3, p_{down}(a) = 0.1 - \text{ if node } a \text{ is even};$$

and node state probabilities are

$$p_{up}(b) = 0.65, p_{mid}(b) = 0.15, p_{down}(b) = 0.2 - \text{ if node } b \text{ is odd.}$$

The *DOWN* state for the network was defined in two ways:

(i) The maximal component has less than $L_1 = 81$ node;
(ii) The maximal component has less than $L_2 = 51$ node.

The LA can be easily applied to the case when the components subject to failure are the nodes, see [9]. Formally, a closure operation can be applied for the nodes too, but its algorithmic implementation is quite complicated. We applied the modified LA without closure.

Table 2.10 $P(DOWN)$ for grid network with two types of nodes

i	$p_2(a); p_2(b)$	$p_1(a); p_1(b)$	$p_0(a); p_0(b)$	L	$P(DOWN)$	RE	CPU sec
1	0.6; 0.65	0.3; 0.15	0.1; 0.2	80	**0.5349**	0.05 %	256
2	0.6; 0.65	0.3; 0.15	0.1; 0.2	50	**0.0710**	0.2 %	197

The simulation results using the modified LA are presented in Table 2.10.

The number of trajectories in the evolution algorithm was $M = 1,000,000$. Let us note that the % of evolution trajectories ending in *DOWN* state was equal 0.1 % for L_1 and 1 % for L_2.

References

1. Barlow RE, Proschan F (1975) Statistical theory of reliability and life testing. Holt Rinehart and Winston Inc, New York
2. Birnbaum ZW (1969) On the importance of different components in multicomponent system. In: Krishnaiah PR (ed) Multivarite analysis-II. Academic Press, New York. pp. 581–592
3. Elperin T, Gertsbakh IB, Lomonsov MV (1991) Estimation of network reliability using graph evolution models. IEEE Trans Reliab 40(5):572–581
4. Gertsbakh I, Shpungin Y (2009) Models of network reliability: analysis combinatorics and Monte Carlo. CRC Press, Boca Raton
5. Gertsbakh I, Shpungin Y (2011) Network reliability and resilience. Springer briefs in electrical and computer engineering. Springer, Heidelberg
6. Gertsbakh I, Shpungin Y (2012) Combinatorial approach to computing importance indices of coherent systems. Probab Eng Inf Sci 26:117–128
7. Gertsbakh I, Shpungin Y (2012) Stochastic models of network survivability. Qual Technol Quant Manag 9(1):45–58
8. Gertsbakh I, Rubinstein R, Shpungin Y, Vaisman R (2014) Permutational methods for performance analysis of stochastic flow networks. Probab Eng Inf Sci 28:21–38
9. Gertsbakh I, Shpungin Y, Vaisman R (2014) Network reliability Monte Carlo with nodes subject to failure. Int J Performability Eng 10(2):161–170
10. Kroese D, Taimre T, Botev IZ (2011) Handbook of Monte Carlo methods. Wiley, New York
11. Lewis TG (2009) Network science: theory and applications. Wiley, Hoboken
12. Ross S (2007) Introduction to probability models, 9th edn. Academic Press, New York

Chapter 3
Interaction of Networks

Abstract The simplest form of two interacting networks is sharing the same set of nodes by two independent networks. For example, the power supply and water supply networks in the same geographic area share the same set of nodes (houses or residencies). Section 3.1 presents several simple results concerning the size of the set of nodes which receive "full" supply, i.e. are adjacent to edges of both types. Here we use some basic facts from the theory of large random Erdos-Renyi or Poisson networks. Section 3.2 considers a system of two or more finite interacting networks. Here the interaction means that a node v_a of network A delivers "infection" to a randomly chosen node v_B in B which in turn, bounces back and infects another randomly chosen node w_a in network A, and so on. As a result, random number Y of nodes in B gets "infected" and fails. We compute, using D-spectra technique, the *DOWN* probability for network B. This model is generalized to the case of several peripheral networks attacking one "central" network B. In this "attack", some of nodes in B will receive more than one hit. The use of DeMoivre combinatorial formula together with D-spectra technique allows obtaining an expression for network B *DOWN* probability in a close form. Finally, Sect. 3.3 extends the results of Sect. 3.2 to the case when the "central" network is ternary. In that case, we must take into account different node behavior that are hit once or more. It is assumed that a node hit only once changes its state from *up* to *mid*. When this node receives another hit, it turns into *down* and remains in it forever. Network *DOWN* probability for this case can be estimated by an efficient Monte Carlo algorithm.

Keywords Networks with colored links · Giant component · Poisson network · Network Interaction · Star-type configuration · D-spectra · Network *DOWN* probability · Ternary network · DeMoivre formula

I. Gertsbakh et al., *Ternary Networks*, SpringerBriefs in Electrical and Computer Engineering, DOI: 10.1007/978-3-319-06440-6_3, © The Author(s) 2014

Fig. 3.1 A network with two
types (*colors*) of edges. *Bold
nodes* are adjacent to edges of
both types

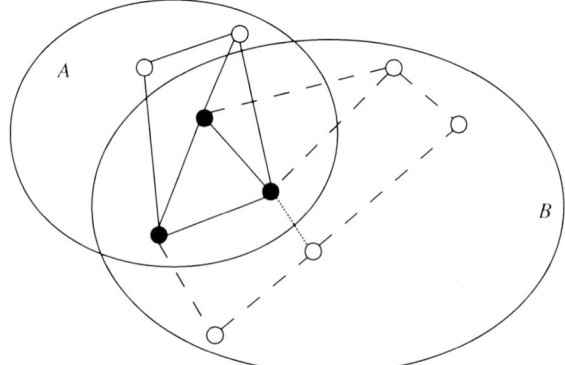

3.1 Networks with Colored Links

We assume that the readers of this section are familiar with basic ideas of network
theory and have read, for example, Chaps. 12–16, of Newman's book "Networks—
An Introduction" [14].

In this section we will consider two networks sharing the same set of nodes. The
first network is $\mathcal{N}_1 = (V, E_1)$, $|V| = n$, $|E_1| = m_1$, and the second network is
$\mathcal{N}_2 = (V, E_2)$, $|E_2| = m_2$.

To make easier the visual perception of these networks we can imagine that the
edges (links) of set E_1 are red, and the links of the second network in set E_2 are
green. On Fig. 3.1 the green links are shown by dotted lines.

A real-life prototype of the above network with two-colored edges are supply
networks: read edges represent, for example, power supply lines, and green edges,
for example, railway roads.

Let us make the following assumptions:

 (i) node set V is large, formally $|V| = n \to \infty$:
 (ii) both networks are random Poisson graphs with average node degrees d_{rd} and
       d_{gr}, correspondingly.
(iii) both networks are stochastically independent.

A network component whose size grows in proportion to network size n as $n \to \infty$
is called *giant component*.

Denote by G_{rd} and G_{gr} the giant components of the "red" and "green" network,
respectively. In supply network context, only the nodes belonging simultaneously
to G_{gr} and G_{rd} receive full supply, i.e. receive both "green" and "red" commodities
(e.g., power and transportation). In our notation the full supply goes only to the nodes
of the set

$$G = G_{gr} \bigcap G_{rd}.$$

Table 3.1 Giant component S as a function of average node degree c

c	1.00	1.25	1.5	1.75	2.0	2.25	2.50	2.75	3.00
S	0	0.371	0.583	0.713	0.797	0.853	0.893	0.920	0.940
c	3.25	3.5	3.75	4.00	4.25	4.5	4.75	5.00	5.25
S	0.955	0.966	0.974	0.980	0.985	0.988	0.991	0.993	0.995
c	5.5	6.0	6.5	7.0	–	–	–	–	–
S	0.996	0.997	0.998	0.999	–	–	–	–	–

Let us call call the set G of nodes adjacent to edges of both colors *rich node set* (RNS). By property (iii)

$$|G| = |G_{gr}| \cdot |G_{rd}|. \tag{3.1}$$

Our first task is to determine the size of G. The following theorem solves the problem.

Theorem 3.1.1 *For a Poisson random graph, the size of the giant component S depends only on the average node degree c via the following equation, see* [14], (12.15):

$$S = 1 - \exp[-c \cdot S]. \tag{3.2}$$

Proof The proof closely follows [14], Sect. 12.5. First note that in Poisson graph the edge between any nodes i and j exists with probability p, and the average node degree equals $c = p(n - 1)$. Denote by u the average fraction of nodes that *do not belong* to giant component. Let us consider node i and find out the probability that it is not connected to giant component. It happens if i is not connected to giant component via any other node. This means that for any other node j either i is not connected to j by an edge, or i is connected to j, but j itself is not in giant component. The first option has probability $1 - p$, and the second option has probability $p \cdot u$. Therefore, the probability of not belonging to giant component via any other of $(n - 1)$ nodes is

$$u = (1 - p + pu)^{n-1} = \left(1 - \frac{c}{n-1}(1 - u)\right)^{n-1}.$$

Setting $n \to \infty$, we obtain from here

$$u = \exp^{-c(1-u)}, \tag{3.3}$$

or replacing $u = 1 - S$, we obtain the desired formula (3.2). #

For any given $c \in (1, \infty)$, the solution of (3.2) can be easily found by **FindRoot** operator of *Mathematica* [17]. Table 3.1 presents a sample of S values for different average node degrees c.

Let us remind that the average number of edges in a network is $m = n \cdot c/2$.

Example 3.1.1 The size of RNS.

Suppose we have a large network with $n = 1000$ nodes. Suppose that the average degree for green and red edges is $d_{gr} = d_{rd} = 2.00$. We see from Table 3.1 that the size of a single-colored component is $S = 0.797$, and the size of RNS is by (3.1) $S_{(2)} = 0.797^2 = 0.635$. Thus RNS has on the average 635 nodes and only these nodes receive "full supply". #

Example 3.1.2 maximal size of RNS built randomly under budget constraint.

Suppose that the cost of a red edge c_{rd} is on the average 5 units, and the cost of the green edge is on the average $c_{gr} = 1$. Suppose, all our budget for designing two-colored network is 9,000 units and suppose that our graph has $n = 1000$ nodes. Each node has the average green degree d_{gr} and the average red degree d_{rd}. Then the total cost of the whole two-colored network equals

$$\frac{c_{rd} d_{rd} n}{2} + \frac{c_{gr} d_{gr} n}{2} = \frac{5 \cdot 1000 \cdot d_{rd}}{2} + \frac{1 \cdot 1000 \cdot d_{gr}}{2},$$

and it should not exceed 9,000. Simplifying, we obtain that:

$$5 d_{rd} + d_{gr} \leq 18. \tag{3.4}$$

Let us find the maximal size of the RNS which is possible to achieve for "green" and "red" Poisson graphs, *within the budget constraint* (3.4). The parameters d_{gr} and d_{rd} must satisfy the constraint:

$$d_{rd} = \frac{18 - d_{gr}}{5} . (*)$$

The numerical procedure for finding the optimal RNS is as follows.

(i) Take $d_{gr} = 1.5$, calculate $|G_{gr}| = S$ from (3.2) using **FindRoot** operator, setting $c := d_{gr}$; calculate d_{rd} from (*) and find $|G_{rd}|$ from (3.2) using **FindRoot** operator and setting $c := d_{rd}$.
(ii) Calculate $G = |G_{rd}| \cdot |G_{gr}|$ and set $d_{rd} := d_{rd} + 0.5$. Return to (i) Stop at the maximal value of G.

The results of this calculations are shown on Fig. 3.2.

   We see that the maximal size $G^* \approx 0.906$ is attained at $c_{gr} = 4$, and $c_{rd} = (18 - 4)/5 = 2.8$. So, the maximal RNS has, on the average, 906 nodes. #

3.2 Interacting Networks

3.2.1 Introduction

Networks attract overgrowing interest in modern research literature. Since network-type systems appear presently in every sphere of our life (social networks, communication networks, supply networks, Internet, etc.), their reliability, survivability

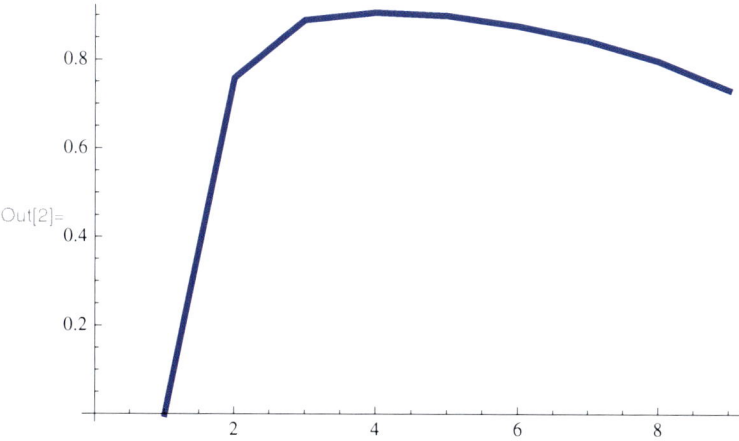

Fig. 3.2 The size of two-colored giant component as a function of c_{gr}

and resilience studies are becoming hot and and important issue. Particular interest is attracted by the so-called cascading failure phenomenon, i.e. massive network failures triggered by failure of a small part of the network components, see e.g. [1, 3, 5, 13].

Majority of the works on network reliability deals with probabilistic failure description and failure modeling within a single network or studies the influence of external "bombardment" of network nodes on its connectivity, or on the size of the giant component, see e.g. [7, 11, 12, 14]. At the same time most modern networks do not exist and do not function in isolated mode. As noted in important paper [3], "…due to technological progress, modern systems are becoming more and more coupled together. While in the past many networks would provide their functionality independently, modern systems depend on one another to provide proper functionality." For example, the financial network of banks and related financial institutions is interconnected with business network of industrial enterprizes. A "failure" of a bank (e.g. bankruptcy) in one country may cause serious problems financing and/or maintaining industrial activity of one or many companies in another part of the world. Similar interdependence exist between power supply network and computer-communication networks since loss of power supply leads to to failures in information supply, and vice versa, i.e. disruption of information flow may cause serious failures in power supply.

There exist a principal difference in reliability studies of very large (formally-infinite) networks and similar studies in finite size (formally-"small") networks. In the studies of very large networks, the main goal is finding the conditions under which the network with n nodes has so-called giant component, i.e. a component of size $O(n)$ as $n \rightarrow \infty$, and establishing the threshold value of network parameters (say, the average node degree) below which the giant component does not exist. Similarly, in studying the development of a cascading failure, and/or an epidemic process, of crucial interest are the conditions which guarantee that the number of damaged (infected) nodes remains limited (formally, is $o(n)$ as $n \to \infty$). As a rule, in the studies of very large networks, the detailed network topology is not introduced and network properties are formulated only in a form of the distribution $\{p_m\}$ of their node degrees, where p_m is a fraction of nodes having degree m.

Contrary to very large networks, the reliability and resilience studies of *finite* networks having 50–200 nodes are carried out under the assumption that we know the detailed network *topology* under study and their failure (*DOWN*) definition. Similarly, the development of cascading failure and network interaction for finite networks is studied in more details that allows to calculate network failure probability and the probabilistic description of failure development.

In this section we will consider network interaction for *binary* networks only. The network components subject to failure will be always the *nodes*. The main assumption will be that if a node v_A of network A fails (gets infected, for example, receives a "blow"), then this node gets *down, and delivers* the infection to a randomly chosen node v_b of network B. After that, the node v_B gets *down*. For ternary networks,which we will consider in the next section, the interaction of nodes is more interesting: if a node v_B of network *B*, receives a single "blow" from some node v_A of network A, then this node changes its state from *up* to *mid*. Upon receiving the second "blow", node v_B changes its state from *mid* to *down*. Subsequent "blows" do not change the state of a node which is already *down*.

3.2.2 Two Interacting Binary Networks

The exposition in this section follows in main lines Sect. 13.2 in [11]. We consider a situation when the cascading failure process is initiated by a *single* node in one of the two interacting networks. The failure of a node in one network causes the failure of another, randomly chosen node in another network, then goes "backwards" to a randomly chosen node of the first network, and so on, until it exhausts itself by hitting a node for a second time. Depending on particular properties of both networks, the process of failure transmission may lead to the failure of one or both interacting networks. Our goal is to find out the distribution of affected nodes and estimate the network failure probability.

We consider networks A and B, having n and m nodes, respectively, $m \geq n$. Nodes of networks are numbered $1_a, 2_a, ..., n_a$ and $1_b, 2_b, ..., m_b$, correspondingly. In further exposition it is convenient to use the terminology of an epidemic process. Initially, at $t = 0$ node 1_a gets infected. It transmits the infection "forward" to a randomly chosen node 1_b. Then 1_b "strikes back" randomly chosen node 2_a. This will be the first turnaround of the process. We may assume that it takes one time unit. It may happen that $1_a \equiv 2_a$ and then the process stops, or, equivalently circulates between 1_a and 1_b. With probability $(1 - 1/n)$ node 1_b delivers infection to another node $2_a \neq 1_a$, and in this case the process goes on in a similar way.

Fig. 3.3 Infection delivery
between networks A and B

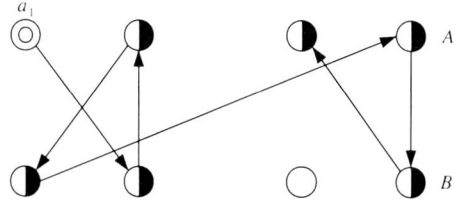

Example 3.2.1 Epidemic transmission between two small networks.
The networks are shown on Fig. 3.3, $n_a = n_b = 4$. Let us find out he probability that
the process will develop by following trajectory

$$\tau = 1_a \mapsto 1_b \mapsto 2_a \mapsto 2_b \mapsto 3_a \mapsto 3_b \mapsto 4_a.$$

It is easy to compute $P(\tau)$:

$$P(\tau) = P(1_a \mapsto 1_b \mapsto 2_a) \cdot P(2_a \mapsto 2_b \mapsto 3_a) \cdot P(3_a \mapsto 3_b \mapsto 4_a) = \frac{3}{4} \cdot (\frac{3}{4} \cdot \frac{2}{4}) \cdot (\frac{2}{4} \cdot \frac{1}{4}).$$

For this example, denote the total number of damaged nodes in A by X_A. $X_A = 1$
if the infection from node 1_b returns to 1_a. Then $P(X_A = 1) = 1/4$. The event
$X_A = 2$ takes place if 1_b hits backward not 1_a, node 2_a delivers infection to any
node except 1_b and 2_b, and 2_b hits one of the already affected nodes in A. This gives
$P(X_A = 2) = (3/4) \cdot (10/16)$. In a similar way, one can establish that

$$P(X_A = 3) = (3/4) \cdot (6/16) \cdot (14/16); \; P(X_A = 4) = (3/4) \cdot (6/16) \cdot (2/16). \#$$

Denote by Y the number of damaged (infected) nodes in B. Let us present without
proof the following

Theorem 3.2.1 [11]
(i) For $r = 3, 4, ..., n$,

$$P(Y = r) = \frac{n-1}{n} \prod_{i=1}^{r-2} \left(\frac{m-i}{m} \cdot \frac{n-i-1}{n} \right) \cdot \frac{m-r+1}{m} \cdot [r/n + (n-r)r/nm]. \quad (3.5)$$

(ii) For $r = 2$,

$$P(Y = 2) = \frac{n-1}{n} \cdot \frac{m-r+1}{m} \cdot (r/n + (n-r)r/nm). \quad (3.6)$$

For $r = 1$, $P(Y = 1) = 1/n + (n-1)/nm$. #
   Since in this section we deal with binary networks, it is necessary to remind
the reader the main tool for the reliability analysis of binary networks with failed
nodes—the cumulative D-spectrum.

3.2.3 Cumulative D-spectrum: A Reminder

This material was explained in detail in [6, 8, 10]. Let us make its brief overview. Let network components subject to failure be numbered as $1, 2, ..., n$. The components may be edges or nodes. In this chapter, we consider only node failures. Node failure means that all edges adjacent to it are erased and the node becomes isolated. Consider a random permutation $\pi = (i_1, i_2, ..., i_n)$ of these numbers. Assign to each permutation probability $1/n!$ Imagine that all components are in *up* state and start turning them *down* moving along π from left to right. After each step of this "destruction" process we check the state of the network and note the number of components needed to be turned *down* to reveal for the first time that the network has failed. Let

$$P(\text{the network failed on step } j) = f_j, \quad j = 1, 2, ..., n.$$

Obviously $\{f_j\}$ is a proper discrete density. In literature it is known also under the name *signature* [15, 16] or *Internal Distribution* (ID) [4]. We will use the cumulative signature $F(x) = \sum_{i=1}^{x} f_i$ which we call *cumulative D-spectrum* or simply D-spectrum ("D" stands for "destruction").

The probabilistic meaning of $F(x)$ is the following: $F(x)$ is the probability that the network is *DOWN* if x of its randomly chosen nodes are *down* (while the remaining $n - r$ nodes are *up*). D-spectrum is a combinatorial invariant of the network, it is completely separated from the random mechanism governing network components failures. For us in this section the following combinatorial formula will be of crucial importance:

$$C(x) = F(x) \cdot \frac{n!}{x!(n-x)!}, \tag{3.7}$$

where $C(x)$ is the number of network failure sets having exactly x components *down* and $n - x$ remaining components *up*.

Using (3.7) we can write the following formula for network *DOWN* probability (we remind that the network has stochastically independent and identical components):

$$P(DOWN) = \sum_{x=1}^{n} C(x)(1-p)^x p^{(n-x)}. \tag{3.8}$$

Here $p = P(\text{component is } up)$.

For our exposition the following formula is of principal importance. Suppose we know the probability $P(Y = x)$ that the number of damaged nodes in our network equals x. Then the total probability that the network is *DOWN* will be

$$P(DOWN) = \sum_{x=1}^{n} P(Y = x) \cdot F(x). \tag{3.9}$$

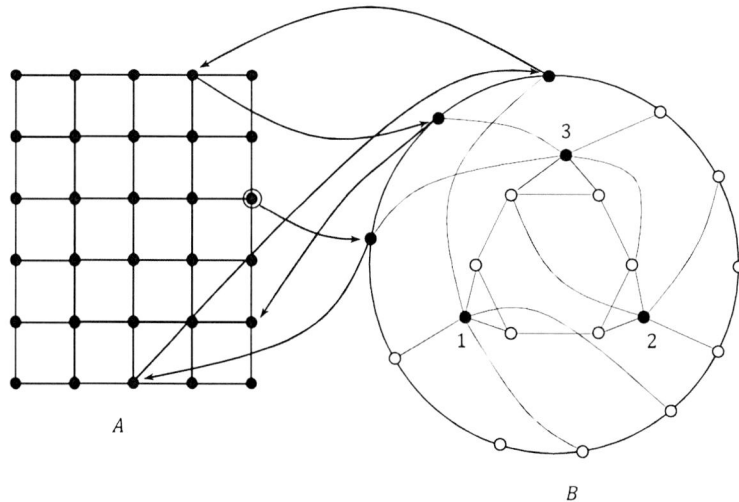

Fig. 3.4 Two interacting networks. Network A initiates the destruction process of B. Nodes 1, 2, 3 are terminals

Table 3.2 $P(Y = x)$ and $F(x)$ for network B

x	$P(Y = x)$	$F(x)$	x	$P(Y = x)$	$F(x)$
1	0.0635	0	11	0.0103	0.5620
2	0.1171	0	12	0.0457	0.7393
3	0.1511	0	13	0.0018	0.8752
4	0.1615	0	14	0.0065	0.9576
5	0.1505	0.0007	15	0.0006	0.9926
6	0.1247	0.0078	16	0.0002	1
7	0.0929	0.0338	17	0	1
8	0.0624	0.0977	18	0	1
9	0.0379	0.2125	19	0	1
10	0.0208	0.3754	20	0	1

Let us illustrate the above theory by an example of two interacting networks.

Example 3.2.3 Two interacting networks
Figure 3.4 presents two small interacting networks [9]. Network A initiates the "infection" process as a result of which some random number of nodes in network B fail. Network B fails if it breaks up into at least two clusters. A cluster is an isolated connected subnetwork which contains at least one special node called terminal. Network B has 20 "regular" nodes, 3 terminal nodes and 32 edges.

Table 3.2 Presents the distribution of the number of nodes hit in network B (columns 3,5) and the D-spectrum of network B (third and sixth column).

Using the above data and formula (3.9) we calculate the probability that network B is *DOWN*:

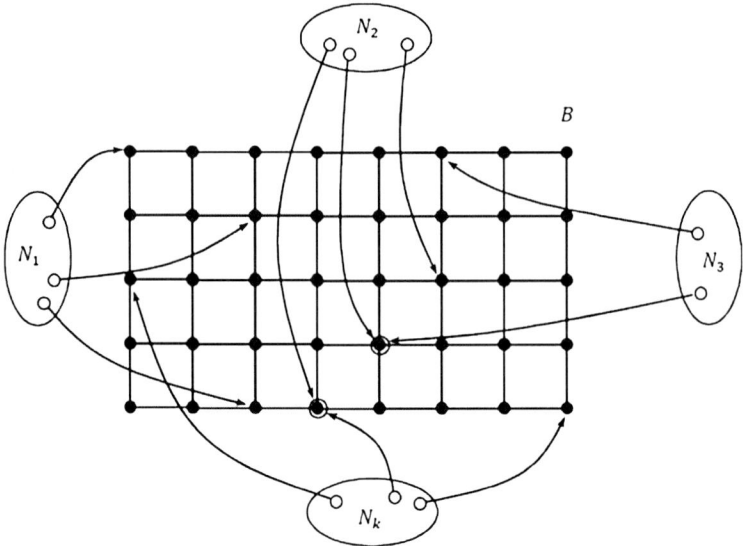

Fig. 3.5 *Circled nodes* received more than one hit

$$P(DOWN) = 0.0754.$$

This probability is rather small because the number of damaged nodes in B is relatively small. #

3.2.4 Star-Type System of Interacting Binary Networks

We consider a collection of "periphery" networks $\mathcal{N}_1, \mathcal{N}_2, \ldots, \mathcal{N}_k$ and one "central" network B, see Fig. 3.5. In the language of epidemic process, each of the n_i nodes of $\mathcal{N}_i$ with probability p_i delivers infection to some randomly chosen node of B. All periphery networks act independently. Each infected node in B fails. The topologies of periphery networks are of no importance, the topology of the central network is known, and its failure state is defined as follows.

To preserve some analogy between the giant component of a very large network and "small" network B, we assume that the network B is *DOWN* if its largest component is greater than some critical value $L_{\min} = 0.5 \cdot n_b$. We remind that node failure means elimination of all edges adjacent to this node. Following [9], we will assume that the number X_i of nodes in B infected (hit) by periphery $\mathcal{N}_i$, $i = 1, \ldots, k$ has a Poisson distribution with parameter $\Lambda_i = n_i \cdot p_i$. This is a realistic assumption if p_i are small and n_i are large. Since all periphery networks act independently, the total number of infection deliveries to center B has Poisson distribution with parameter

Table 3.3 D-spectrum of 5 × 8 grid B

x	$F(x)$	x	$F(x)$
1	0	11	0.32229
3	0	12	0.56158
4	0	13	0.83440
5	0.00066	14	1.00000
6	0.00311	15	1.00000
7	0.01106	16	1.00000
8	0.03053	17	1.00000
9	0.07416	18	1.00000
10	0.16402	19-40	1.00000

$$\Lambda = \sum_{i=1}^{k} \Lambda_i \cdot p_i,$$

and

$$P(X = r) = e^{-\Lambda} \frac{\Lambda^r}{r!}, \quad r = 0, 1, 2, \dots$$

By our assumption, a node hit by infection that originated in some "periphery" network gets *down*. Quite often there will be a situation that a particular node of B will be hit more than once, even by the same periphery network. Of course, such node will remain *down*. To calculate the probability that B is *DOWN* we need to know the distribution of nodes which are hit at *least once*.

This situation reminds the famous model considered in combinatorics and known as "occupancy" problem [2], p. 242. In this model, r balls are randomly thrown into m identical boxes. The probability that exactly k boxes will contain at least one ball is given by the famous DeMoivre formula:

$$p(k|r) = \frac{m!}{k!(m-k)!} \sum_{t=0}^{k} (-1)^t \frac{k!}{t!(k-t)!} \left(\frac{k-t}{m}\right)^r, k = 1, \dots, \min(m, r). \quad (3.10)$$

Example 3.5 Failure probability of the central network.
Assume that the periphery networks create a Poisson flow with total intensity $\Lambda = 20$, i.e. the average number of hits is 20. Network B is a rectangular 5 × 8 grid. Its D-spectrum is presented in Table 3.3.

Now everything is ready to compute *DOWN* probability of network B. The probability that B is *DOWN given* that it has received r hits from periphery is equal

$$P_B(DOWN|r) = \sum_{j=1}^{\min(m,r)} p(j|r) \cdot F(j). \quad (3.11)$$

Then

$$P_B(DOWN) = \sum_{r \geq 1} P(X = r) \cdot P_B(DOWN|r). \qquad (3.12)$$

Using the above formulas we find out that

$$P_B(DOWN) = \mathbf{0.891}.\#$$

3.3 Star-Type System with Central Ternary Network

In this section we consider a system of interacting networks consisting of a "central" network B and a collection of "periphery" networks. This system is similar to the one considered in the previous section, but with one principal difference: the central network B is *ternary* and *not* binary. The components of B subject to failure are *nodes*. They can be in three states: *up,* intermediate (*mid*) and *down*. For example, a node which is in *mid* transmits information only to a part of links adjacent to it, i.e. it is partially isolated. A node which is *down* is completely isolated.

Each node of periphery network $\mathcal{N}_i$ fails with probability p_i, as a result of which some randomly and independently chosen node in B receives a "hit". Physically, it may correspond to a disruption of power or information supply, infection delivery, loss of communication, etc. If a node b in B receives a single "hit", it changes its state from *up* to *mid*. After the second hit, with no matter of its origin, the node gets *down* and remains in this state forever.

As in the previous section, we assume that the total flow of "hits" from periphery to the center creates Poisson flow with intensity Λ. It means that the probability that the network B receives K hits equals

$$P(K) = e^{-\Lambda} \frac{\Lambda^K}{K!}, \quad K = 0, 1, 2, \dots$$

If B has m nodes, then the situation here reminds the classical occupancy problem: K balls are randomly and independently thrown into m boxes. There is, however, a significant difference with respect to the previous model. We are interested not only in the number of boxes which are nonempty, but we must know also the number D of boxes containing two or more balls, and the number of boxes S containing exactly one ball. (Denote by E the number of empty boxes: $E = m - S - D$).

Our task is to calculate the probability that the central network will be *DOWN*. Let us remind the probabilistic meaning of the ternary D-spectrum. $F_r(x)$ is the probability that the network is *DOWN* if r randomly chosen nodes are *up* and x nodes are *down*. Suppose that we know that $E = r$ and $D = x$. Then $F_E(D)$ is the probability that the network is *DOWN*. Unfortunately, we don't have a formula analogous to DeMoivre's which would give us the *joint* distribution of (S, D). Suppose we are able to find the joint conditional density $\Psi_K(r, x) = P(E = r, D = x|K)$. Then the formula for system *DOWN* probability would be the following:

$$P_B(DOWN) = \sum_{K=0}^{\infty} e^{-\Lambda} \frac{\Lambda^K}{K!} \sum_{r \geq 0, k \geq 0: r+k \leq m} \Psi(r, x|K) \cdot F_r(x). \qquad (3.13)$$

In practical calculations, we will approximate of $P(DOWN)$ by using the following straightforward Monte Carlo procedure.

Algorithm: *Estimation of P(DOWN) for central Ternary Network*
(**0**) It is assumed that the ternary D-spectrum $F_r(x)$ is known. Set Counter **P(Down)** := 0;
(**i**) Generate K—Poisson random variable with parameter Λ.
(**ii**) Locate randomly and independently K balls into m boxes, where m is the number of nodes in the central network;
((**iii**) Count D—the number of boxes containing two or more balls;
((**iv**)) Count E—the number of empty boxes; put $r := E$; $x := D$;
(**v**) Set **P(Down)** := **P(Down)** $+ F_r(x)$;
Repeat steps (**i**)–(**v**) M times;
(**vi**) Estimate $P(DOWN)$ by $\widehat{P}(DOWN) := \mathbf{P(Down)}/M$.

References

1. Buldyrev SV, Parshani R, Stanley HF, Havlin S (2010) Catastrophic cascade of failures in interdependent networks. Nature 464:1025–1028
2. David FN, Barton DE (1962) Combinatorial chance. Charles Griffin and Co Ltd, London
3. Dickinson M, Havlin S (2012) Epidemics on interconnected networks. Phys Rev E85:066109
4. Elperin T, Gertsbakh IB (1991) Estimation of network reliability using graph evolution models. IEEE Trans Reliab 40(5):572–581
5. Gao J et al (2012) Robustness of network formed by n interdependent networks with one-to-one correspondence between nodes. Phys Rev E85:066134
6. Gertsbakh I, Shpungin Y (2009) Models of network reliability: analysis combinatorics and Monte Carlo. CRC Press, Boca Raton
7. Gertsbakh I, Shpungin Y (2011) Network reliability and resilience. Springer briefs in electrical and computer engineering. Springer, Berlin
8. Gertsbakh I, Shpungin Y (2012) Combinatorial approach to computing importance indices of coherent systems. Probab Eng Inf Sci 26:117–128
9. Gertsbakh I, Shpungin Y (2012) Failure development in a system of two connected networks. Transp Commun 13(4):255–260
10. Gertsbakh I, Shpungin Y (2012) Stochastic models of network survivability. Qual Technol Quant Manag 9(1):45–58
11. Gertsbakh I, Shpungin Y (2014) Single-source epidemic process in a system of two interconnected networks. In: Frenkel I, Lisniansky A, Karagrigoriu A, Kleyner A (eds) Chapter 13, in Applied reliability engineering and risk analysis. Wiley, New York
12. Lewis TG (2009) Network science. Theory and applications. John Wiley & Sons Inc, New York
13. Li W, Bashan A et al (2012) Cascading failures in interconnected lattice networks. Phys Rev Lett 108(22):228702

14. Newman MEJ (2010) Networks: an introduction. Oxford University Press, New York
15. Samaniego FJ (1985) On closure under ifr formation of coherent systems. IEEE Trans Reliab 34:69–72
16. Samaniego FJ (2007) System signatures and their application in engineering reliability. Springer, New York
17. Wolfram S (1991) Mathematica: a system for doing mathematics by computer, 2nd edn. Addison-Wesley Publishing Company, New York

Index

I. Gertsbakh et al., *Ternary Networks*, SpringerBriefs in Electrical
and Computer Engineering, DOI: 10.1007/978-3-319-06440-6,
© The Author(s) 2014

Printed by Publishers' Graphics LLC
LMO140508.23.35.75